ESTIMATING THE INCIDENCE OF RAPE AND SEXUAL ASSAULT

Panel on Measuring Rape and Sexual Assault in
Bureau of Justice Statistics Household Surveys

Candace Kruttschnitt, William D. Kalsbeek, and Carol C. House, *Editors*

Committee on National Statistics

Division of Behavioral and Social Sciences and Education

NATIONAL RESEARCH COUNCIL
OF THE NATIONAL ACADEMIES

THE NATIONAL ACADEMIES PRESS
Washington, D.C.
www.nap.edu

THE NATIONAL ACADEMIES PRESS 500 Fifth Street, NW Washington, DC 20001

NOTICE: The project that is the subject of this report was approved by the Governing Board of the National Research Council, whose members are drawn from the councils of the National Academy of Sciences, the National Academy of Engineering, and the Institute of Medicine. The members of the committee responsible for the report were chosen for their special competences and with regard for appropriate balance.

This study was supported by the Bureau of Justice Statistics, U.S. Department of Justice, via grant number SES-1024012 between the National Academy of Sciences and the National Science Foundation. Support for the work of the Committee on National Statistics is provided by a consortium of federal agencies through a grant from the National Science Foundation (grant number SES-1024012). Any opinions, findings, conclusions, or recommendations expressed in this publication are those of the author(s) and do not necessarily reflect the views of the organizations or agencies that provided support for the project.

International Standard Book Number-13: 978-0-309-29737-0
International Standard Book Number-10: 0-309-29737-0
Library of Congress Control Number: 2014934964

Additional copies of this report are available from the National Academies Press, 500 Fifth Street, NW, Keck 360, Washington, DC 20001; (800) 624-6242 or (202) 334-3313; http://www.nap.edu.

Printed in the United States of America

Suggested citation: National Research Council. (2014). *Estimating the Incidence of Rape and Sexual Assault.* Panel on Measuring Rape and Sexual Assault in Bureau of Justice Statistics Household Surveys, C. Kruttschnitt, W.D. Kalsbeek, and C.C. House, Editors. Committee on National Statistics, Division of Behavioral and Social Sciences and Education. Washington, DC: The National Academies Press.

THE NATIONAL ACADEMIES

Advisers to the Nation on Science, Engineering, and Medicine

The **National Academy of Sciences** is a private, nonprofit, self-perpetuating society of distinguished scholars engaged in scientific and engineering research, dedicated to the furtherance of science and technology and to their use for the general welfare. Upon the authority of the charter granted to it by the Congress in 1863, the Academy has a mandate that requires it to advise the federal government on scientific and technical matters. Dr. Ralph J. Cicerone is president of the National Academy of Sciences.

The **National Academy of Engineering** was established in 1964, under the charter of the National Academy of Sciences, as a parallel organization of outstanding engineers. It is autonomous in its administration and in the selection of its members, sharing with the National Academy of Sciences the responsibility for advising the federal government. The National Academy of Engineering also sponsors engineering programs aimed at meeting national needs, encourages education and research, and recognizes the superior achievements of engineers. Dr. C. D. Mote, Jr., is president of the National Academy of Engineering.

The **Institute of Medicine** was established in 1970 by the National Academy of Sciences to secure the services of eminent members of appropriate professions in the examination of policy matters pertaining to the health of the public. The Institute acts under the responsibility given to the National Academy of Sciences by its congressional charter to be an adviser to the federal government and, upon its own initiative, to identify issues of medical care, research, and education. Dr. Harvey V. Fineberg is president of the Institute of Medicine.

The **National Research Council** was organized by the National Academy of Sciences in 1916 to associate the broad community of science and technology with the Academy's purposes of furthering knowledge and advising the federal government. Functioning in accordance with general policies determined by the Academy, the Council has become the principal operating agency of both the National Academy of Sciences and the National Academy of Engineering in providing services to the government, the public, and the scientific and engineering communities. The Council is administered jointly by both Academies and the Institute of Medicine. Dr. Ralph J. Cicerone and Dr. C. D. Mote, Jr., are chair and vice chair, respectively, of the National Research Council.

www.national-academies.org

PANEL ON MEASURING RAPE AND SEXUAL ASSAULT IN BUREAU OF JUSTICE STATISTICS HOUSEHOLD SURVEYS

CANDACE KRUTTSCHNITT (*Cochair*), Department of Sociology, University of Toronto
WILLIAM D. KALSBEEK (*Cochair*), Department of Biostatistics, University of North Carolina at Chapel Hill
PAUL P. BIEMER, RTI International and the Odum Institute, University of North Carolina at Chapel Hill
JOHN BOYLE, ICF International, Rockville, MD
BONNIE S. FISHER, School of Criminal Justice, University of Cincinnati
KAREN HEIMER, Department of Sociology, University of Iowa
LINDA LEDRAY, SANE-SART Resource Service, Minneapolis, MN
COLIN LOFTIN, School of Criminal Justice, University at Albany, State University of New York
RUTH D. PETERSON, Department of Sociology (emerita) and Criminal Justice Research Center, Ohio State University
NORA CATE SCHAEFFER, Department of Sociology, University of Wisconsin–Madison
TOM W. SMITH, NORC at the University of Chicago
BRUCE D. SPENCER, Department of Statistics and Institute for Policy Research, Northwestern University

CAROL C. HOUSE, *Study Director*
NANCY KIRKENDALL, *Senior Program Officer*
DANIEL L. CORK, *Senior Program Officer*
ESHA SINHA, *Associate Program Officer*
AGNES GASKIN, *Administrative Assistant*

COMMITTEE ON NATIONAL STATISTICS
2012-2013

LAWRENCE D. BROWN (*Chair*), Department of Statistics, The Wharton School, University of Pennsylvania
JOHN M. ABOWD, School of Industrial and Labor Relations, Cornell University
DAVID CARD, Department of Economics, University of California, Berkeley
ALICIA CARRIQUIRY, Department of Statistics, Iowa State University
CONSTANTINE GATSONIS, Center for Statistical Sciences, Brown University
JAMES S. HOUSE, Survey Research Center, Institute for Social Research, University of Michigan, Ann Arbor
MICHAEL HOUT, Department of Sociology, New York University
SALLIE ANN KELLER, Social and Decision Analytics Lab, Virginia Bioinformatics Institute at Virginia Tech, Arlington, VA
LISA LYNCH, The Heller School for Social Policy and Management, Brandeis University
SALLY C. MORTON, Department of Biostatistics, Graduate School of Public Health, University of Pittsburgh
RUTH D. PETERSON, Criminal Justice Research Center, Ohio State University
EDWARD H. SHORTLIFFE, Department of Biomedical Informatics, Columbia University and Arizona State University
HAL S. STERN, Donald Bren School of Information and Computer Sciences, University of California, Irvine
JOHN R. THOMPSON, NORC at the University of Chicago*
ROGER TOURANGEAU, Westat, Inc., Rockville, MD

CONSTANCE F. CITRO, *Director*
JACQUELINE R. SOVDE, *Program Associate*

*Resigned August 2013.

Acknowledgments

The Panel on Measuring Rape and Sexual Assault in Bureau of Justice Statistics Household Surveys was convened in March 2011 by the Committee on National Statistics (CNSTAT), with input from the Committee on Law and Justice, under the auspices of the Division of Behavioral and Social Sciences and Education of the National Research Council (NRC). The study sponsor, the Bureau of Justice Statistics (BJS) of the U.S. Department of Justice, charged the panel to recommend the best methods for obtaining national statistics on rape and sexual assault on an ongoing basis for the noninstitutionalized population of the United States in conjunction with the BJS household surveys.

The panel held five in-person meetings to organize its work, gather information, prepare recommendations, and finalize its report. To achieve these goals, the panel relied on a wide range of individuals both within and outside of the NRC and BJS. First, we must acknowledge the outstanding contributions of our fellow panel members who brought expertise in survey design, including questionnaire design and interview mode, the National Crime Victimization Survey, and the broader areas of rape and sexual assault and related policies and programs. The panel offers special thanks to Bonnie Fisher and Karen Heimer for their work on Appendix D, and to William Kalsbeek for his work on Appendix E. See Appendix F for biographical sketches of the panel members and project staff. As part of our task, the panel was asked to communicate with the agency contracted by BJS—Westat—to pilot two survey designs that would compare alternative measures of rape and sexual assault. We are grateful to both then BJS director, James Lynch, and BJS senior statistical advisor, Allen Beck, for facilitating our communications with Westat and for the openness and camaraderie shown to us by the director of the Westat project, David Cantor.

The panel convened two workshops (December 2011 and June 2012) to gather information from experts on police reports and self-reports of rape and sexual assaults, from consumers of these statistics, and from experts on statistical methods for assessing bias in the estimates of rates of sensitive and rare events. We are indebted to the participants of these two workshops who shared their invaluable knowledge with the panel. (See Appendix B for the workshop agendas and participants.) Some of the individuals who presented at these workshops also prepared written papers, which expanded on the topics they discussed at the workshops. Their papers were very valuable to the panel as we deliberated and prepared our recommendations, and these individuals deserve additional thanks: Ronet Bachman; Jim Chromy and David Wilson; Janet Lauritsen; Ken Rasinski; and Carol Tracy, Jennifer Long, Terry Fromson, and Charlene Whitman.

This report was reviewed in draft form by individuals chosen for their diverse perspectives and technical expertise, in accordance with procedures approved by the NRC Report Review Committee. The purpose of this independent review is to provide candid and critical comments that will assist the institution in making its published report as sound as possible and to ensure that the report meets institutional standards for objectivity, evidence, and responsiveness to the study charge. The review comments and draft manuscript remain confidential to protect the integrity of the deliberative process.

We wish to thank the following individuals for their review of this report: Lynn A. Addington, Department of Justice, Law and Society, American University; Kathleen M. Brown, School of Nursing, University of Pennsylvania; Dean G. Kilpatrick, National Crime Victims Research and Treatment Center, Department of Psychiatry and Behavioral Sciences, Medical University of South Carolina; Virginia M. Lesser, Survey Research Center, Department of Statistics, Oregon State University; Nancy A. Mathiowetz, Department of Sociology, University of Wisconsin–Milwaukee; Elizabeth Stasny, Department of Statistics, Ohio State University; Patricia Tjaden, Tjaden Research Corporation, Denver, Colorado; Roger Tourangeau, Executive Office, Westat, Rockville, Maryland; and Min Xie, School of Criminology and Criminal Justice, Arizona State University.

Although the reviewers listed above have provided many constructive comments and suggestions, they were not asked to endorse the conclusions or recommendations nor did they see the final draft of the report before its release. The review of this report was overseen by Richard A. Kulka, independent consultant, Raleigh, North Carolina, and Linda McCauley, Nell Hodgson Woodruff School of Nursing, Emory University. Appointed by the NRC's Report Review Committee, they were responsible for making certain that an independent examination of this report was carried out in accordance with institutional procedures and that all review comments were carefully considered. Responsibility for the final content of this report rests entirely with the authoring panel and the institution. However, the panel gratefully acknowledges that its report is more complete and its points made more clearly because of questions asked and suggestions made by the individuals who participated in this review.

The panel also recognizes that its work was greatly facilitated by many NRC staff members. Connie Citro, director of CNSTAT, attended all of our meetings and workshops and provided invaluable expertise throughout the duration of our deliberations. Daniel Cork, senior program officer with CNSTAT, also provided us with critical information that allowed us to link our work with the broader assessment of BJS that was conducted in 2008 and 2009. Esha Sinha, associate program officer, prepared early drafts of several sections of the report. Nancy Kirkendall, senior program officer with CNSTAT, assisted the panel in arranging for several commissioned papers.

The panel's study director, Carol House, was invaluable to the panel. From the outset, she was a critical conduit to Westat and BJS and outside reports and information essential to the panel's deliberations. In addition to keeping the panel's work on track, Carol's own professional background in survey statistics enabled her to capably serve as a bridge to connect the various expertise among the panel members.

The panel also wants to thank the three other members of the CNSTAT staff who worked with the panel to facilitate its meetings and workshops: Agnes Gaskin, Alicia Jaramillo-Underwood, and Anthony Mann.

Finally we recognize the many federal agencies that support CNSTAT directly and through a grant from the National Science Foundation. Without their support and commitment to improving the national statistical system, the panel work that is the basis of this report would not have been possible.

Candace Kruttschnitt and William D. Kalsbeek, *Cochairs*
Panel on Measuring Rape and Sexual Assault in
Bureau of Justice Statistics Household Surveys

Contents

SUMMARY 1

1 INTRODUCTION 15
Study Context, 16
Panel Charge and Activities, 17
Report Overview, 20

2 LEGAL DEFINITIONS AND CONTEXT 23
Historical Context, 23
Overview of State Statutes, 25
Conclusion, 33

3 DATA FROM LAW ENFORCEMENT AGENCIES 35
Police Reports, 35
UCR Reports, 40

4 NATIONAL CRIME VICTIMIZATION SURVEY 49
History and Development, 50
Target Population and Sample Design, 52
Data Collection and Survey Mode, 59
Survey Instruments, 59
Estimation and Products, 63

5 SELECTED OTHER SURVEYS ON RAPE AND SEXUAL ASSAULT 71
National Women's Study (1989-1991), 72
National Violence Against Women Study (1995-1996), 75
National College Women Sexual Victimization Study (1997), 78
National Intimate Partner and Sexual Violence Survey (2010), 84

6 COMPARISON OF RAPE AND SEXUAL ASSAULT ACROSS DATA SOURCES 91
Definitions Used, 92
Survey Context, 95
Target Population, Sampling Frames, and Sample Size, 96
Data Collection Mode and Response Rates, 99
Measures of Rape and Sexual Assault, 102

7 POTENTIAL SOURCES OF ERROR IN THE NCVS: SAMPLING, FRAME, AND PROCESSING 109
Sampling Error, 110
Frame Error, 122
Processing Error, 124

8 POTENTIAL SOURCES OF ERROR: NONRESPONSE, SPECIFICATION, AND MEASUREMENT 127
Nonresponse Error, 127
Specification Error, 136
Measurement Error, 138

9 SYNOPSIS OF POTENTIAL ERRORS IN THE NATIONAL CRIME VICTIMIZATION SURVEY 153
Potential Sources of Error: Summary, 154
Obstacles to High-Quality Estimates, 155
General Recommendations, 158

10 NEW DIRECTIONS FOR MEASURING RAPE AND SEXUAL ASSAULT 161
A New Survey to Measure Rape and Sexual Assault, 162
Specialized Training and Monitoring, 175
Research, 176
Communication with Data Users, 177
Wrap-Up, 180

REFERENCES 183

APPENDIXES

A Need for the Study 193
B Workshop and Public Meetings: Agendas and Participants 197
C Links to Questionnaires of the National Crime Victimization Survey 205
D Selected Surveys Measuring Rape: An Overview 207
E Statistical Rationale Behind Some Initial Findings on the Relative Statistical Plausibility of a Multiple-Frame Approach to Estimating the Victimization Rate of Rape and Sexual Assault 247
F Biographical Sketches of Panel Members and Staff 257

Summary

Rape and sexual assault are highly injurious victimizations, and accurate information about them is difficult to obtain because they are seriously underreported to law enforcement. The National Crime Victimization Survey (NCVS) measures these victimization rates,[1] along with details on the victims, as part of its overall mission of measuring all criminal victimizations. However, data users have expressed concern that rape and sexual assault appear to be undercounted on the NCVS.

The Bureau of Justice Statistics (BJS), which has responsibility for the NCVS, has committed to a multiyear project to better understand the reasons for the possible underestimation of rape and sexual assault on the NCVS. As part of this effort, BJS asked the National Research Council, through its Committee on National Statistics, to convene an expert panel

[1]Throughout the report, we use two specific terms to discuss victimization rates of rape and sexual assault: incidence and prevalence. The *incidence rate* refers to the measure of the total number of incidents (or events) that occurred in a given period. It counts the total number of incidents or victimizations; it does not count the number of individual victims. In epidemiology, this rate is often referred to as the "event rate." Incidence rates are generally calculated over a specific time period, such as 12 months. The *prevalence rate* refers to the number of victims. It counts the number of individuals who have been victimized at least once; it does not count the total number of incidents. Thus, a lifetime prevalence rate measures the number of individuals who have been raped or sexually assaulted at least once in their lifetimes. A 12-month prevalence rate would measure the number of individuals who had been raped or sexually assaulted at least once in that 12-month period. In epidemiology, the term *incidence rate* is often used to measure the number of "first time events," which is what we are calling the prevalence rate.

to investigate these issues and recommend best practices for measuring rape and sexual assault on the NCVS and other BJS household surveys.

There are two quite different perspectives for the measurement of rape and sexual assault—the criminal justice perspective and the public health perspective. These different perspectives have led to methodological differences in designing and implementing surveys, which, in turn, have resulted in different estimates of the incidence rates. The NCVS reflects the criminal justice perspective, and its purpose is to measure criminal victimizations: "point-in-time" events that are judged to be criminal. In contrast, surveys that reflect the public health perspective look at victimization as a condition that endures over a period of time, and may not necessarily be criminal. These surveys are less focused on identifying point-in-time events.

The panel was formally charged to "assess the quality and relevance of statistics on rape and sexual assault from the NCVS and other surveys contracted for by other federal agencies as well as surveys conducted by private organizations," examining issues such as the "legal definitions in use by the states for these crimes, best methods for representing the definitions in survey instruments so that their meaning is clear to respondents, and best methods for obtaining as complete reporting as possible of these crimes in surveys, including methods whereby respondents may report anonymously." Thus, the panel took a fresh look at the problem of measuring incidents of rape and sexual assault from the criminal justice perspective, but the panel was not constrained to fit this measurement within the NCVS or to restrict its recommendations to specific methodologies that BJS has used in the past.

MEASUREMENT

The first part of this report focuses on methodology and vehicles used to measure rape and sexual assault. Looking first at legal definitions for these crimes, the panel found that there are considerable differences across jurisdictions. The differences include basic terminology, the level of "force" required before the victimization becomes criminal, and the concepts of "lack of consent" and the "capacity to consent." Chapter 2 of this report provides details of what the panel learned about these legal definitions.

Along with these differences, the measurement of rape and sexual assault has been implemented in different venues, in different ways, using different definitions and different methodologies. The result has been different levels in the estimates. Chapters 3 through 6 in this report describe these different venues, with highlights below.

The Federal Bureau of Investigation (FBI) annually publishes *Crime in the United States*, which includes statistics from the Uniform Crime Reports (UCR) Summary Reporting System (SRS). The UCR SRS is based on monthly crime counts, by type of crime, from approximately 17,000 law enforcement agencies around the country.

Until 2013, the FBI directed law enforcement agencies to report rape crimes using a restrictive definition established in 1929, "carnal knowledge of a female forcibly and against her will," with a further explanation that "carnal knowledge" is penetration (however slight) of a penis into a vagina. Although a new definition for rape was established in 2013 for purposes of reporting in the UCR that is better aligned with current state and federal laws, there are still questions about whether the UCR can accurately capture all kinds of rape and sexual assault incidents. Two major concerns are that rape and sexual assault are underreported to law enforcement and sometimes downgraded by police.

The NCVS was established in part to provide another source of crime statistics beyond those supplied by police reports. It is a national household survey with the goal of obtaining information on a broad set of criminal victimizations (including rape and sexual assault) from the victims rather than law enforcement. The Census Bureau conducts it on an ongoing basis for BJS. The NCVS selects housing units through a multistage design that uses the infrastructure built for the decennial census. Individuals (12 years of age and older) residing at the selected housing units are interviewed. In 2011, the NCVS had reports from approximately 143,000 household members. Each address remains in the sample for 3 years, with interviews every 6 months.

BJS has established its own definitions of rape and sexual assault for estimation with the NCVS, broader than the one previously used by the FBI. In the NCVS, respondents are not asked to judge whether or not a crime has taken place but to report incidents in a number of categories. The NCVS estimated 217,331 rapes and sexual assaults in 2011 in the United States.

Users of NCVS data have expressed concern about potential underestimation of rape and sexual assault on the NCVS, in part because a number of other surveys have measured higher levels of those victimizations. Those independent surveys include (but are not limited to) the National Women's Study (1989-1991); the National Violence Against Women Study (1995-1996); the National College Women Sexual Victimization Study (1997); and the National Intimate Partner and Sexual Violence Survey (2010). The surveys differ in many ways, including the definition used for rape; context in which data are collected; target population, sampling frame, and sample size; and data collection mode, response rates, and adjustments for nonresponse. Given these many differences, it is not surprising that the resulting estimates of rape and sexual assault are substantially different.

The panel found that a comparison across these sources of estimates of rape was particularly problematic because of the differences in the populations targeted, the definitions used, the data collection methodology, and the survey timing. The panel determined that it could not scientifically conclude which source was overall better, and it does not recommend any source as the best or as a standard. However, in reviewing all of this material, the panel judges that it is likely that the NCVS is undercounting rape and sexual assault victimization.

ASSESSMENT OF THE NCVS

All surveys are subject to errors, and the NCVS is no exception. An assessment of the errors and potential errors in a survey is important to understanding the overall quality of the estimates from that survey and to initiate improvements. Total survey error is a concept that involves a holistic view of all potential errors in a survey program, including both sampling error and various forms of nonsampling error.

The panel undertook an examination of the total error structure of the NCVS with the intent of identifying areas that were particularly problematic and that could contribute to underestimation of rape and sexual assault. This review of potential sources of error covered sampling error, frame error, processing error, nonresponse error, specification error, and measurement error. The panel also assessed the training and monitoring of interviewers for the NCVS.

Although it has identified areas where errors seem likely to occur, the panel, with limited time and resources, was not able to conduct a complete error profile of the NCVS. Such an error profile would measure the actual level of error and its impact on the estimation of rape and sexual assault. The panel encourages BJS to conduct an in-depth total error profile of the NCVS, to update and complete its documentation, and to make its research agenda and results more accessible and transparent to the public (Recommendations 9-1, 9-2, and 9-3 in Chapter 9).

It is important to note that the panel did not perform the same in-depth examination of the error structure of the other surveys for measuring rape and sexual assault because of limitations of time and resources. Presenting findings focused on the NCVS does not imply that the panel believes that the other surveys have fewer errors: the panel did not examine them carefully and so cannot draw overall conclusions about their error structures.

Sampling Error

The target population of the NCVS is the noninstitutionalized population of the United States, 12 years of age and older. One measure of sampling error, the coefficients of variation (CVs) for the number of rape and sexual assault victimizations, is approximately 14 percent at the national level, with considerable year-to-year variation. Furthermore, the sampling errors for estimates of important subpopulations are quite large. As a result, BJS does not provide estimates for rape and sexual assault for those subpopulations, instead providing estimates only for the more aggregated category of serious violent crimes.

> **CONCLUSION 7-1 The National Crime Victimization Survey, which is designed as an omnibus victimization survey, is efficient in measuring the many types of criminal victimizations across the United States, but it does not measure the low incidence events of rape and sexual assault with the precision needed for policy and research purposes. Comparisons across subgroups and years are particularly problematic.**

BJS made a major methodological change in 2011 in how the NCVS handles "series victimization." Series victimization is defined as when a single respondent reports six or more separate but similar criminal victimizations over the reference period but is unable to recall each event individually or describe each one in detail to the interviewer. The old methodology suppressed these reports of multiple victimizations, contributing to the underestimation of rape and sexual assault. In 2011, the agency changed these procedures to count the number of reported victimizations in the series up to a maximum of 10. From a statistical point of view, the new series victimization procedures give the weighted outliers a very large impact on the estimates. The effect of this change increased the estimates of the incidence rate of rape and sexual assault by 55 percent and created more year-to-year fluctuations.

> **CONCLUSION 7-2 Records identified as series victimizations create an outlier problem in the estimation process for the National Crime Victimization Survey. The current method for handling series victimization, though an improvement over the method used until 2011, allows these relatively rare reports to have a large impact on the national estimates of rape and sexual assault and creates large year-to-year volatility.**

Frame and Processing Error

Conducting a major household survey is complex. In the NCVS, as in any major survey, errors may arise in the process of constructing, maintaining, or sampling from a frame. All large household surveys, including the NCVS, use complex processes to edit, summarize, and publish data and the calculated estimates. The NCVS also includes a complex process for the classification of each reported victimization by type of crime. The panel concludes that these survey processes contribute to errors in the NCVS estimates and that there is a lack of transparency in the edit and processing procedures (see Conclusions 7-3, 7-4, and 7-5 in Chapter 7).

Response Error

With regard to survey response, the NCVS has maintained a moderately high level of overall survey response. In 2011, the person-level response rate[2] was 88 percent. These response rates have decreased several percentage points over the previous decade, but not substantially. A major question with regard to response rates is whether the nonresponse on the NCVS causes a bias in the estimates. Even though one major analysis of potential bias in the NCVS in 2009 found little evidence for nonresponse bias, the panel has some reservations and concludes that the NCVS may have a nonresponse bias related to estimates of sexual victimization.

CONCLUSION 8-1 The overall unit response rates, as calculated, on the National Crime Victimization Survey are moderately high and have been reasonably stable over the past 10 years. Although an independent analysis concluded that the methods that the Bureau of Justice Statistics uses to adjust for nonresponse appear to provide a satisfactory correction for nonresponse bias at the unit level, our panel has reservations about that analysis and remains concerned that there may be a nonresponse bias related to sexual victimization.

Because the NCVS is a panel survey with seven waves of data collection over 3 years, the panel was concerned about panel attrition: whether household members were less likely to respond or more likely to completely drop out over time. Because BJS does not provide NCVS response rates by wave, the panel calculated unweighted person-level response rates by time-in-sample. We found that response over the survey's seven waves differs by important subgroups. In particular, younger individuals participate in fewer waves, as do individuals who did not live as a "couple." These results are of concern because some of the people (younger people and females not living as part of a couple) who participate less appear to be people who are more at risk for being victims of rape and sexual assault.

CONCLUSION 8-2 There appears to be notable panel attrition over the 3 years in the National Crime Victimization Survey. This attrition is particularly problematic for estimating rape and sexual assault because

---

[2]The person-level response rate is the percentage of household members in cooperating households who responded on an individual questionnaire.

some people at greater risk for being victimized by these crimes—young people and females not living as part of a couple—are also some of those most likely to drop out before the seven waves of the NCVS have been completed.

CONCLUSION 8-3 Although the Bureau of Justice Statistics publishes annual response rates for the National Crime Victimization Survey (NCVS), the published data do not include important details of response, such as mode of data collection and attrition rate. Such details are needed by data users for a thorough assessment of the quality of NCVS estimates.

Item nonresponse occurs when a respondent completes a substantial portion of a questionnaire (enough to count the interview as "complete") but does not provide answers to certain key items. The panel's opinion is that item "refusals" to questions about sexual victimization are difficult to identify. If a respondent does not want to report a rape or sexual assault, then he or she is more likely to answer that he or she was not victimized rather than by directly refusing to answer the question. Thus, what may be an item refusal is most likely counted as just a "no."

CONCLUSION 8-4 The panel believes it is likely that item refusals on questions about sexual victimization on the National Crime Victimization Survey may be recorded as if they were a "no" response rather than item nonresponse when a respondent does not want to report a victimization. Another possibility is for a respondent to sometimes answer "no" on screening questions to avoid additional questions in the survey.

Specification Error

Specification error may occur when there is a mismatch between what the survey is measuring and what it is intended to measure. A critical concept for the NCVS is to identify whether and when a respondent has experienced a rape or sexual assault. However, the complex, multifaceted definitions of what is meant by rape and sexual assault are translated into a few simple words in the omnibus screening questionnaire such as rape, attempted rape, other type of sexual attack, and unwanted sexual activity. These words do not convey the complexity of the intended concepts.

CONCLUSION 8-5 There is serious specification error in the National Crime Victimization Survey measurement of rape and sexual assault. Although the Bureau of Justice Statistics has developed clear defini-

tions of the concepts, they are replaced in the omnibus screener by ambiguous wording that does not convey the multifaceted concepts to respondents.

Measurement Error

Measurement error includes a large family of errors that may occur when a response on a survey results in the collection of inaccurate or incomplete information. The panel identified three characteristics of the current NCVS procedures that foster measurement errors: the use of ambiguous terms, such as *rape*, in questions (see above), the overall context (crime) in which questions are asked, and the lack of privacy in responding to the survey's questions.

CONCLUSION 8-6 Words, such as "rape" and "sexual assault," on the National Crime Victimization Survey may not be consistently understood by survey respondents. Other surveys have used more behaviorally specific words to describe a specific set of actions. More specific wording of questions would be understood more consistently by all respondents and thus lead to more complete and accurate answers.

CONCLUSION 8-7 Questions about incidents of rape and sexual assault in the National Crime Victimization Survey are asked in the context of a criminal victimization survey and embedded within individual questions that describe other types of crimes. This context may inhibit reporting of incidents that the respondent does not think of as criminal, did not report to the police, or does not want to report to police.

CONCLUSION 8-8 The current data collection mode and methods of the National Crime Victimization Survey do not provide adequate privacy for collecting information on rape and sexual assault. This lack of privacy may be a major reason for underreporting of such incidents.

As part of examining measurement error, the panel identified several problems with the training provided to interviewers on the NCVS and the subsequent monitoring of the interview process. For 10 years, until recently, there was no refresher training for interviewers, and the reinstated training offered only limited focus on the special training needs for sensitive questions about sexual victimizations. The training on these questions is not reinforced through the day-to-day survey process because of the low incidence of such reports. In addition, there is inadequate monitoring of the field data collection processes, of both in-person and decentralized telephone interviewing by field representatives, to ensure consistent quality (see Conclusions 8-9 and 8-10 in Chapter 8).

SUMMARY AND RECOMMENDATIONS

With respect to measuring rape and sexual assault, the panel identified four major barriers for quality measurement on the NCVS:

1. a sample design that is inefficient for measuring these low incidence events,
2. the context of "crime" that defines the survey,
3. a lack of privacy for respondents in completing the survey, and
4. the use of words with ambiguous meaning for key measures in the questionnaire.

The first three barriers are intrinsic to the basic structure and processes of the NCVS, which appear to work well for measuring other criminal victimizations. These barriers cannot be overcome by making modifications to the NCVS without potentially compromising the quality of the NCVS design for measuring other crime rates. Only the last one—use of ambiguous terms—could be readily addressed within the structure and operations of the current NCVS.

CONCLUSION 10-1 The best methods for measuring rape and sexual assault cannot be implemented without separating that measurement from the measurement of other criminal victimizations.

A New Survey and Improved Methods

RECOMMENDATION 10-1 The Bureau of Justice Statistics should develop an independent survey—separate from the National Crime Victimization Survey—for measuring rape and sexual assault.

The panel makes several recommendations regarding the design and implementation of the independent survey, including a multiple frame approach (see Chapter 10). They are adaptations to the sampling and measurement strategies currently in place for NCVS and focus on finding ways to isolate and strategically oversample segments of the population where the risk of rape and sexual assault victimization is relatively greater.

RECOMMENDATION 10-2 The Bureau of Justice Statistics should rigorously compare the relative cost-efficiency of alternative sample

designs for the recommended new survey to measure rape and sexual assault, including the multiple frame approach described in Chapter 10.

RECOMMENDATION 10-3 The Bureau of Justice Statistics should continue to publish annual estimates of rape and sexual assault criminal victimizations, using the recommended new survey to do so. However, if that is not possible, then the Bureau should conduct the recommended new survey on a fixed schedule, such as every 2 or 3 years, and use data from both the National Crime Victimization Survey and the new survey to calculate annual estimates of rape and sexual assault.

The panel supports the multiwave structure of the NCVS for the proposed stand-alone survey, and it endorses the use of bounded recall procedures to control telescoping effects. However, the panel stresses that more can be learned through research into ways to improve the quality of data obtained using bounded recall. In addition, the panel has serious concerns about the current adjustments to wave 1 data to compensate for potential telescoping.

RECOMMENDATION 10-4 The recommended new survey should have a longitudinal structure with at least two waves to allow the use of bounded recall. Research should be conducted to determine an optimal length of reference period specifically for reporting rape and sexual assault victimizations. The Bureau of Justice Statistics should reassess the methodology used to adjust for forward telescoping if data from the bounding interview are used in estimation.

The context in which survey questions are asked is a critical element in obtaining accurate responses in any survey; it is particularly critical for questions about incidents of rape and sexual assault. The panel believes that framing these questions within a criminal context limits accurate responses. In addition, the panel strongly supports the wording of survey questions so that they describe specific actions rather than the more ambiguous term "rape."

RECOMMENDATION 10-5 The questionnaire and protocols for the recommended new survey should have a neutral context, such as a health survey. The Bureau of Justice Statistics should explore several neutral alternatives while continuing to use both a victimization screening questionnaire and an incident report. The questions on both of these instruments should be reworded to incorporate behaviorally specific questions.

The lack of respondent privacy in the interview setting is a critically important barrier to obtaining truthful response in screening survey respondents for rape and sexual assault victimization. Self-administration of the screener instrument and incident report by a single randomly chosen member of participating households is the best strategy for overcoming this barrier. Because there are other survey error implications to this approach (e.g., on sampling error), BJS needs to carefully consider the relative merits of this approach for obtaining the most accurate possible victimization rates.

RECOMMENDATION 10-6 The recommended new survey should be conducted in a self-administered mode. The wave 1 contact would involve a personal visit and audio computer-assisted self-administered interviewing technology. Only one individual in each selected household should be selected for this survey to increase the respondent's privacy.

The panel believes that people who have been victimized while not having the capacity to consent to sexual actions may be undiscovered by the current survey instruments. This oversight should be directly addressed by expanding the conceptual definition of rape and sexual assault to explicitly include these victims and by adding questions that probe in this area.

RECOMMENDATION 10-7 The Bureau of Justice Statistics' definition of rape and sexual assault should be expanded to include victimizations when the victim does not have the capacity to consent to the sexual actions of the offender.

The exact process of determining the "type of crime" classification of an incident reported on the NCVS based on responses provided by the respondent is not clearly portrayed for data users.

RECOMMENDATION 10-8 The Bureau of Justice Statistics should make more transparent the link between responses on the survey screener and incident reports and the final type of crime classification of those incidents of potential rape and sexual assault.

The NCVS produces estimates of victimization rates, but it should also enable data analysts to identify important predictors of victimization. A generous array of social-demographic covariates obtained for each respondent is needed to more fully realize this potential.

RECOMMENDATION 10-9 The recommended new survey should include a number of covariates to add to the richness of the dataset for analysis. The Bureau of Justice Statistics should hold an expert-user workshop as it develops the new survey. A major purpose of the workshop would be to obtain advice on the covariates that could best improve the usefulness of the dataset for research, advocacy, and policy purposes.

The current procedures for handling series victimizations create estimation problems for the NCVS, resulting in large increases in the estimates of rape and sexual assault and their accompanying standard errors, and much greater year-to-year fluctuation in the estimates.

RECOMMENDATION 10-10 The Bureau of Justice Statistics should reassess the methodological change made to the National Crime Victimization Survey regarding series victimization and should investigate alternative procedures that are more effective in respect to measurements of rape and sexual assault. This reassessment should involve formal input by experts on outlier adjustment techniques and by data users who can help assess the relative tradeoffs in quality.

Training

There are some straightforward steps that could be taken to improve the quality of training and supervision being offered to the Census Bureau field representatives who administer the NCVS, particularly as they relate to the sensitive nature of the topic of rape and sexual assault. The monitoring of the process can also be improved.

RECOMMENDATION 10-11 The Census Bureau and the Bureau of Justice Statistics should provide specialized training for field representatives on how to assist the respondents and answer questions on the sensitive subjects of rape and sexual assault. The interaction between respondents and field representatives should be recorded using computer-assisted recorded interviewing technology.

Ongoing Research Program

The panel recommends that BJS develop an ongoing program of research addressing a variety of design-related problems related to the recommendations mentioned previously. There are 11 key research topics to investigate (see Chapter 10):

1. the cost-efficiency of introducing disproportionate stratified sampling of those at higher risk for rape and sexual assault victimization;
2. the cost-efficiency of supplementing the standard area household sampling frame with one or more frames derived from administrative sources with higher concentrations of victims (e.g., college residence hall records, police files, emergency room records, etc.);
3. the best estimation approach to deal with telescoping effects arising from the use of bounded questions in a longitudinal setting;
4. the effect of changing the survey to have a more neutral context;
5. the effects of following a neutral/behavioral orientation for questions used to screen for rape and sexual assault victimization;
6. the joint sampling and measurement error implications of self-administration of a single respondent chosen in each participating household;
7. the effect of expanding the definition of rape and sexual assault to include those without the capacity to give their consent to the offender;
8. the error and cost implications of improved training and supervision of field representatives;
9. ways to improve estimation in the presence of series victimization;
10. effective models to estimate the underreporting of rape and sexual assault on the NCVS based on data from a periodic independent survey; and
11. issues related to collecting data on rape and sexual assault criminal victimization from adolescents (12-17 years of age) because of their relatively high risk of that victimization.

The panel recommends that this research be conducted in a coordinated manner because many of the issues to be investigated interrelate.

> **RECOMMENDATION 10-12 The Bureau of Justice Statistics should conduct a series of coordinated research investigations to enable it to resolve each of the preceding specific issues in developing the design for the recommended stand-alone survey on rape and sexual assault.**

Enhanced Communication with Data Users

As BJS moves in new directions to improve its measurement of rape and sexual assault, the agency needs to embrace external advice from the data user and statistical communities, and set up mechanisms to ensure open, regular, and effective communication with these communities. We offer three recommendations to BJS to better enhance this communication: establishment of a permanent advisory committee, regular data user conferences, and steps for major methodological changes (see Recommendations 10-13, 10-14, 10-15 in Chapter 10).

The panel applauds BJS for its openness in addressing ways to fulfill its important mission to provide estimates of rape and sexual assault and is confident that the analyses and recommendations in this report can contribute to improvements in measuring these injurious victimizations.

1

Introduction

The crimes of rape and sexual assault are among the most injurious that perpetrators can inflict on other individuals. These crimes are devastating, extending beyond the initial victimization to such consequences as unwanted pregnancy, sexually transmitted infections, post-traumatic stress disorder, depression, flashbacks, sleep disorders, eating disorders, post-incident substance abuse, self-harm, and even suicide (Rape, Abuse and Incest National Network, n.d.). The effects are often long lasting and can lead to health and work productivity issues for years. Using analyses from three different studies of the cost of crime, Heaton (2010) estimated that the victim-related costs[1] for nonlethal rape and sexual assault are between $150,000 and $283,626 per victim.

Understanding the frequency and context under which rape and sexual assault are committed is vital in directing law enforcement and victim-support resources. These data can influence public policy in the areas of public health, mental health, and education. They also can be used to identify and implement interventions that will reduce the risk of future victimizations. Unfortunately, accurate information about the extent of rape and sexual assault is particularly difficult to obtain because these crimes are seriously underreported to law enforcement.

[1]Total victim-related costs include both tangible and intangible costs. The tangible costs include such things as medical treatment and lost wages. Intangible costs are larger and much harder to estimate. They include such things as lost quality of life resulting from fear and psychological effects of victimization.

STUDY CONTEXT

In *Ensuring the Quality, Credibility, and Relevance of U.S. Justice Statistics*, the National Research Council (NRC) made a strong recommendation to the Bureau of Justice Statistics (BJS) regarding its mission: "[It] must ensure that the nation has quality annual estimates of levels and changes in criminal victimization" (National Research Council, 2009, p. 10). A major tool in accomplishing this mission is the National Crime Victimization Survey (NCVS), an ongoing BJS survey of the noninstitutional population of the United States to estimate victimization rates[2] by type of crime, along with details about the victims and the social context of those victimizations. The NCVS provides an independent source of information on criminal victimization—a source of data for the calibrating of police-reported incidents and an indicator of the extent that victimization incidents go unreported to law enforcement. "Since the survey began full-scale data collection in the early 1970s, the NCVS has become a major social indicator for the United States. Serving as a complement to the official measures of crime reported to the police . . . the NCVS has been the basis for better understanding the cost and context of criminal victimization" (National Research Council, 2008, p. 2).

The NCVS is an omnibus victimization survey. As such, it has a broad mandate and focus to include a wide array of different types of victimizations, including both crimes against people and crimes against property. This is a difficult task, and approaches that may be overall best for a general survey may be less optimal for measuring a specific type of victimization.

This report focuses on a narrow portion of the NCVS: the estimation of incidence rates for rape and sexual assault. There is controversy as to whether the NCVS is providing accurate estimates of the rates of rape and sexual assault, in part because some other sources of statistics over the past two decades have shown higher levels of victimization than those estimated through the NCVS and its predecessor, the National Crime Survey.

A 1992 report by the National Victim Center in collaboration with the Crime Victims Research and Treatment Center, *Rape in America, A Report to the Nation* (Kilpatrick, Edmunds, and Seymour, 1992), based on the National Women's Study (NWS), estimated that 683,000 adult women were raped during a 12-month period in 1989 and 1990. That study estimated that only 16 percent of those rapes were reported to the police. For a similar time period, the Uniform Crime Reporting (UCR) Program of the Federal Bureau of Investigation (FBI) reported 102,500 rape victims (a figure that is quite similar to the 16 percent of rapes reported to police in the NWS).[3]

---

[2]See footnote 1 in the Summary.

[3]The UCR compiles counts of incidents of crimes from participating law enforcement agencies and so is necessarily limited to counting crimes reported to the police.

The NCVS (then known as the National Crime Survey [NCS] prior to its redesign) estimated 130,000 victims for approximately the same time period. Thus, the data from the NWS indicated a level of victimization that was more than five times that reported in the NCVS.[4]

More recently, the Centers for Disease Control and Prevention launched a major survey on the subject, the National Intimate Partner Violence and Sexual Violence Survey (NISVS). This survey has a number of objectives, one of which is to provide estimates of the extent of rape and sexual assault. In 2010, it counted 1,270,000 victims of rape and attempted rape,[5] many times higher than the NCVS estimate of victimizations (188,380) that same year (Bachman, 2012; Black et al., 2011; Bureau of Justice Statistics, 2011).

Thus, an important barrier to understanding the extent of rape and sexual assault in the United States is the existence of multiple sources of information providing different answers. These surveys and programs have somewhat different objectives, are conducted within a different "context," and use different measurement tools. The end result is that they provide different estimates of the extent of rape and sexual assault. This in turn, creates confusion for the public, for law enforcement, for policy makers, for researchers, and for victim advocacy groups.

PANEL CHARGE AND ACTIVITIES

BJS has committed to a multiyear project to better understand the magnitude and reasons for potential underreporting of rape and sexual assault on the NCVS and the reasons for the differences between NCVS estimates and data on rape and sexual assault from other sources. As part of this effort, BJS asked the Committee on National Statistics of the NRC to convene an expert panel to investigate these issues and recommend best practices for measuring the incidents of rape and sexual assault from BJS household surveys and to assess the quality and relevance of NCVS statistics on rape and sexual assault. The Panel on Measuring Rape and Sexual Assault in Bureau of Justice Statistics Household Surveys was appointed to carry out the task. The panel was asked to examine such issues as legal definitions of rape and sexual assault and best methods for operationalizing these definitions in survey instruments so that they are understood by respondents; to examine the factors that affect underreporting and propose solutions that can minimize that problem; and to look at ways that can ensure that

---

[4]The quality, methods, and definitions used in the two surveys differed, and the NCS has since been redesigned as the current NCVS.

[5]The NISVS has a target population of males and females 18 years or older. It has a broader definition of rape than does the NCVS.

BOX 1-1
Statement of Task

An ad hoc panel will examine conceptual and methodological issues surrounding survey statistics on rape and sexual assault and recommend to the Bureau of Justice Statistics best methods for obtaining such statistics on an ongoing basis. The panel will assess the quality and relevance of statistics on rape and sexual assault from the National Crime Victimization Survey and other surveys contracted for by other federal agencies as well as surveys conducted by private organizations. Issues to be examined include legal definitions in use by the states for these crimes, best methods for representing the definitions in survey instruments so that their meaning is clear to respondents, and best methods for obtaining as complete reporting as possible of these crimes in surveys, including methods whereby respondents may report anonymously. The panel will organize a workshop and commission papers as principal means of gathering information to support its deliberations. The panel will issue a report with its findings and recommendations at the conclusion of a 21-month study. The panel's scope of work will not include surveys in nonhousehold, institutional settings, such as prisons.

respondents are able to report privately and anonymously on the NCVS. The formal charge to the panel is shown in Box 1-1.

In addressing the panel at its initial meeting, James P. Lynch, then the director of BJS, provided background for the panel's work. His statement is presented in Appendix A. He stressed the importance of having crime statistics that are generated independently of, and in addition to, those statistics provided through police reports. He said that he believes that this is particularly important for rape and sexual assault "since there is good evidence that the majority of these offenses are not reported to the police. These offenses remain the darkest of the 'dark figure' of crime." The National Research Council (2008) and others (Baumer and Lauritsen, 2010; Lauritsen and Heimer, 2008) have also highlighted the importance of the NCVS. They have found that no other national data on criminal victimization are collected with this level of detail and published annually.

Addressing rape and sexual assault specifically, Lynch said that two approaches have evolved to measuring these crimes through surveys, with different conceptual definitions and methodologies. Not surprisingly, these two approaches provide different estimates of victimization, which in turn provides confusion for the public. Lynch explained:

> One group emphasizes the *criminal justice perspective* and the other takes a *public health approach*. The criminal justice school emphasizes crime as a point in time event and employs legal definitions (but plain language

> descriptions) of the target behavior. . . . The public health approach emphasizes victimization as a condition that endures over time and requires treatment to restore the victim. Consequently, there is less concern with identifying point-in-time events that may comprise the condition and legal definitions are of less concern than commonly understood definitions of the behavior.

In broad terms, the NCVS represents the criminal justice perspective, and the NISVS and other surveys described in this report represent the public health perspective.

Lynch expressed several specific expectations for the panel:

- Take a fresh look at the problem, drawing from what the criminal justice and public health schools have done but not being held captive by these traditions. The principal goal of the panel is to consider a wide range of alternative self-report survey designs to measure the incidence and prevalence of the crimes of rape and sexual assault and to recommend an optimum design.
- Recommend whether this optimum design can be incorporated into the ongoing NCVS program and, if so, how.
- Finally, to work closely with Westat in field testing the redesign options.

Thus, Lynch said he hoped for considerable flexibility from the panel in looking for "best practices." He asked the panel to take a fresh look at this problem, drawing from approaches that have been successful in both the criminal justice and public health approaches. He noted that he does not assume that the optimum design could be incorporated into the ongoing NCVS program.

The panel addressed its charge in two main phases, with the initial phase to gather relevant information. It completed a careful assessment of the design, implementation, and output from various administrative and survey data systems that have been used to produce estimates of rape and sexual assault in the United States. Outside investigators doing work in related areas were commissioned to prepare papers for a public workshop held in June 2012. (Agendas for the workshop and public meetings of the panel are presented in Appendix B.) The panel also consulted *Surveying Victims: Options for Conducting the National Crime Victimization Survey* (National Research Council, 2008) to enhance its understanding of issues and to identify any recommendations from that study that are relevant today for measuring rape and sexual assault. Beyond a review of available documentation, this phase included a field exercise in which several panel members participated as respondents in mock interviews on the NCVS (criminal justice perspective) and the NISVS (public health approach).

In phase two, the panel developed a prioritized list of ideas that had potential to improve the quality of survey estimates of rape and sexual assault. These ideas reflected all components of total survey error, including imperfect sampling frames, inefficiency in sample selection, nonparticipation by sampled households and individuals, misspecification and other measurement problems, and processing errors. From these ideas, the panel developed the key elements of an "optimum" design for the measurement of rape and sexual assault. Some features of the recommended design closely resemble the current design of the NCVS, while others would require BJS to move away from the NCVS for implementation. The panel's recommendations provide BJS with specific guidance on key aspects of this design.

As requested by Director Lynch, the panel worked publicly with investigators at Westat, which had been contracted by BJS to develop a pilot project to test two alternative survey designs to measure rape and sexual assault. Westat staff presented the status of their work at each of the panel's open meetings and participated in open discussion at those meetings with panel members and other participants. Following the June 5-6, 2012, public workshop, several panel members provided individual informal comments to Westat on the draft plans that Westat presented at that public workshop. They are provided for the purpose of full disclosure in the Public Access File (see http://www8.nationalacademies.org/cp/information.aspx?key=Internet_FAQ [March 2014]). The Westat team and the panel kept each other advised of their project timelines for various activities throughout the process. Following NRC policy, there was no sharing of the panel's deliberations, conclusions, and recommendations with Westat or BJS during this study.

Note that it was beyond the charge of this panel to compare estimates from the NCVS for other types (beyond rape and sexual assault) of victimizations to examine whether the NCVS possibly underestimated or overestimated these victimizations. The NRC has completed other studies of the broader NCVS (National Research Council, 2008, 2009).

REPORT OVERVIEW

The panel is charged with recommending best practices for measuring the incidences of criminal victimization of rape and sexual assault through BJS household surveys. This charge was not intended to nor did it restrict the panel to only considering solutions within the structure of the NCVS. A critical first step is to establish a broad understanding of the conceptual definitions for the terms *rape* and *sexual assault*. A second step is to look at how these concepts are measured operationally in existing data systems (including, but not limited to, the NCVS)—the methods used and the results obtained.

In pursuit of these goals, Chapters 2 through 6 of the report first consider definitions of rape and sexual assault and their legal histories and then detail several important sources of data on these victimizations.

In order to design a national survey on rape and sexual assault, consistent definitions for these criminal victimizations have to be defined. This task is complicated by the fact that these crimes are generally based on state, rather than federal, statutes. Chapter 2 explores rape and sexual assault in a legal context, analyzing the components of existing legal definitions and their differences and commonalities across jurisdictions. It also covers the historical context from which modern laws against rape and sexual assault have evolved and the changes in those statutes over time. The purpose of this chapter is to look for commonalities across jurisdictions that would be important to include in operational definitions.

The next four chapters detail the data that are available and the methods used to obtain them. Chapter 3 describes the statistical information about crimes available from law enforcement agencies. It looks at police incident reports and the FBI's UCR. This system provides the official measure of crimes reported to the police, and it thus provides an important baseline for comparing other sources of survey-based data.

Addressing the estimation of victimization from population surveys, Chapter 4 provides a description of the NCVS. It includes a review of the survey's history, methodology, and implementation, as well as the survey's resulting estimates of rape and sexual assault. Chapter 5 looks at four other important surveys of rape and sexual assault that have been conducted over the past 25 years. They have used different methods and produced different results.

Chapter 6 compares and contrasts the data discussed in the previous three chapters, focusing on methods and results. It offers the panel's conclusions that inform the central part of our charge to propose improvements to the design of BJS household surveys that measure rape and sexual assault.

In the second half (Chapters 7 through 10) of the report the panel turns to in-depth analyses of the NCVS and its adequacy for the goal of accurately measuring rape and sexual assault. It is important to note that the report does not provide the same in-depth evaluation of the other sources of data described in Chapter 5. Chapters 7, 8, and 9 focus on the NCVS as the current vehicle through which BJS measures victimization rates for rape and sexual assault. This focus on the NCVS reflects a prioritization of the panel's time and resources: it does not imply that the panel believes that the other sources have better measures of rape and sexual assault or are subject to fewer errors.

Chapters 7 and 8 cover the error structure for the NCVS: sampling error, specification error, imperfect sampling frames, nonresponse, measurement error, and error in data processing. For each of these error types, the panel has evaluated the potential to generate errors in estimates of rape and sexual assault so that solutions could be identified.

Chapter 9 summarizes the analyses from Chapters 7 and 8, clarifying which potential errors may have the largest effects on the reported estimates. The chapter specifically identifies four major obstacles for accurately estimating incidents of rape and sexual assault in the current NCVS, which are the basis for the panel's conclusions and recommendations in Chapter 10. The chapter also includes four recommendations for BJS.

Chapter 10 details the panel's conclusion that the NCVS is not an adequate vehicle for the goal of accurate measurement and presents the panel's recommendations for best practices, including a recommendation for a separate survey to measure rape and sexual assault victimizations. It provides guidelines on the optimum design of this new survey, as well as lower-cost variations. It also includes recommendations for specialized training and monitoring, research, and enhanced communication with data users.

Of special note, this report uses the terms "low incidence" and "statistically rare" to describe the criminal victimizations of rape and sexual assault because the frequency with which they occur makes them difficult to measure in a household population survey. The report explores a number of statistical practices that better measure rare attributes in a population. The panel's position is that these victimizations are critically important to measure accurately and the terms "low incidence" or "statistically rare" do not diminish that importance.

2

Legal Definitions and Context

For any national survey measuring rape and sexual assault victimizations, uniform definitions of those victimizations are needed. Because the Bureau of Justice Statistics (BJS) focuses specifically on criminal victimization, these definitions need to conform as much as possible to existing legal definitions. Because the crimes of rape and sexual assault fall mostly under state rather than federal criminal statutes and these statutes are not uniform across jurisdictions, this presents an immediate difficulty. The panel thus decided to review the statutes of 50 states and U.S. territories, examine their differences, and extract the commonalities that may be important to include in an operational definition for BJS. For this task, the panel was able to draw on the work of the Women's Law Project and AEquitas[1] (AEquitas, 2012; Tracy et al., 2012), whose reports are the major source of the information in this chapter.

HISTORICAL CONTEXT

Historical context has greatly influenced the formation and enforcement of current laws regarding rape and sexual assault, and any review of these laws and definitions must begin by examining that context. Tracy et al. (2012, p. 1) note:

[1]The Women's Law Project is a Pennsylvania-based nonprofit law firm that worked with the Federal Bureau of Investigation (FBI) to develop an updated definition of rape in 2012. AEquitas is a nonprofit organization that provides prosecutors with support, training, mentorship, and resources in cases involving violence against women.

> Rape and sexual assault laws are complex and evolving. Rape originated as a crime against property, not a crime against a person. As such, the crime related to patriarchal inheritance rights and a female's reproductive capacity and therefore was limited to a crime against unmarried virgins and included only forcible penile/vaginal penetration. These laws have evolved but retain vestiges of their archaic origins. The result is inconsistency and variability in sex crime terminology and elements from state to state as well as anomalies.

American jurisprudence was developed on these foundations, creating the basic elements of laws regarding rape. As laws developed, the Model Penal Code, first promulgated in 1962 by the American Law Institute, provided state legislatures with the then best thinking on how to develop criminal codes. The model code defines rape as "sexual intercourse with a female not his wife" by force or threat of severe harm (Model Penal Code, 1980, as cited in Tracy et al., 2012, p. 5). To be a felony of the first degree, the code says that the rape must be accompanied by serious bodily harm and that the victim and offender could not have been social companions or had a history of sexual activity.

Coupled with these basic elements were procedural requirements that are not included for other types of assault, and they appear to have been based on the belief that women lie about being raped:

- a need for a prompt complaint to police,
- a requirement for independent corroboration of the victim's testimony or evidence of serious injury, and
- the admittance of testimony regarding the victim's sexual history and character.

This basic approach to state laws regarding rape and sexual assault began to change in earnest in the 1970s, partially in response to feminist activism (Belknap, 2001). As Tracy et al. (2012, p. 6) write:

> As a result of this activism, most states have expanded the definitions of sex crimes to eliminate disparities based on gender and marital status. They have also rescinded the requirements of resistance, corroboration, and reporting requirements and prohibited introduction of a woman's past sexual history. It is now well established that penetration of orifices other than the vagina is a felony. Issues of force and consent continue to change but clear trends in the evolution of the law are identifiable. The definition of force is broadening beyond overt physical force alone to include other modes of coercion. There is an increasing recognition that penetration without consent or any additional force beyond penetration is a serious sexual offense.

Even with the changes that have been made since the 1970s, the current laws on sex crime do not reflect the dynamics of rape and sexual assault as they actually occur. In the 2010 National Intimate Partner and Sexual Violence Survey (NISVS), Black et al. (2011) found that

- the majority of both female and male victims knew their offenders,
- most rapes do not involve physical force or use of a weapon, and
- rape does not generally result in serious physical injury other than the rape itself.

The legal history and underlying beliefs about sex crimes have continued to influence the way police handle reports of these crimes and how prosecutors pursue these cases (Brunson and Miller, 2006; Tracy et al., 2012). That constellation of laws and beliefs contributes to the lack of willingness by many victims to report the crime to law enforcement (Carbone-Lopez, Slocum, and Kruttschnitt, forthcoming; Felson and Pare, 2005; Meloy and Miller, 2011). A special government report (Bureau of Justice Statistics, 2012b) estimated that only 35 percent of these crimes are reported to the police. Kilpatrick, Edmunds, and Seymour (1992) estimated an even lower percentage of police reports, between 16 and 33 percent.

The underreporting of the crimes of rape and sexual assault to law enforcement is one of the basic reasons that other sources of information, such as data from the National Crime Victimization Survey, are important. However, some of the same fears and feelings of shame and self-blame remain barriers to victims' reporting of rape and sexual assault incidents on surveys (Rasinski, 2012; Weiss, 2010).

OVERVIEW OF STATE STATUTES

Understanding and measuring the crimes of rape and sexual assault are difficult because statutes related to these crimes differ considerably among the 50 states, the District of Columbia, U.S. territories, federal jurisdictions, and the Uniform Code of Military Justice. AEquitas (2012) provides a very detailed representation for each jurisdiction in its publication. This section attempts to look at several essential elements of those statutes and how they interrelate with each other and with other criminal statutes.

Tracy et al. (2012, p. 14) identify the following essential elements, how they are handled in state statutes, and compares and contrasts them in statutes:

- penetration, contact without penetration, and noncontact exposure;
- use of force;
- absence of consent;

- victim's capacity to consent; and
- whether the conduct was for the purpose of sexual arousal or degradation.

These elements are discussed separately below. The last section of the chapter reviews how different combinations affect decisions about criminality and its severity.

Essential Elements

Penetration, Contact Without Penetration, and Noncontact Exposure

The various statutes that describe penetration crimes use different terminology: rape, sexual assault, sexual abuse, sexual battery, carnal knowledge, sexual intercourse, sexual penetration, sexual act, deviate sexual assault, etc. Any specific term may have a different meaning in a different jurisdiction. Because of these differences, one has to question the use of any of these terms on a survey questionnaire.

Jurisdictions also differ as to whether the description of penetration includes the quantifier of "slight" or "however slight." There is usually a description of the object that inflicts the penetration (penis, tongue, fingers, other objects) and the part of the victim's body that is penetrated (vagina, anus, mouth, etc.). Slight variations in the combinations of elements may make a difference as to whether the offense is "criminal" and in the severity of potential punishment.

Penetration by itself is not unlawful. It becomes unlawful if coupled by force, without consent, or if the victim does not have the capacity to consent when the penetration occurs. However, different jurisdictions have differing requirements for these accompanying elements before an act is classified as a crime.

The definition of nonpenetration contact includes incidents that involve touching or fondling the intimate parts of another person. New Mexico requires that such contact be skin to skin, but other jurisdictions include touching through clothing. In some jurisdictions, these crimes may also include urinating or defecating on a person for sexual arousal or degradation (see Table 2-1).

The definition of noncontact exposure involves "forced" viewing of the offender's body parts or sexual activity, such as exposing one's genitals in a public place (see Table 2-1).

TABLE 2-1 Contact and Exposure Crimes with Requirements of Sexual Arousal, Gratification, Degradation, or Humiliation

Crime	Requirement of Sexual Arousal or Gratification	Requirement of Degradation or Humiliation
Indecent contact	Alabama, California, Colorado, Connecticut, District of Columbia, Georgia, Illinois, Indiana, Maine, Maryland, Michigan, Minnesota, Missouri, Montana, Nebraska, New Hampshire, New Jersey, New York, North Carolina, North Dakota, Ohio, Oregon, Pennsylvania, Rhode Island, South Dakota, Tennessee, Texas, Utah, Virginia, Washington, West Virginia, Wisconsin, Wyoming, American Samoa, Guam, Virgin Islands, federal law,* Uniform Code of Military Justice	Connecticut, District of Columbia, Montana, New Hampshire, New Jersey, New York, Ohio, South Dakota, Wisconsin, federal law,* Uniform Code of Military Justice
Indecent exposure	**Alabama**, Alaska, **Colorado, Connecticut, Illinois, Indiana, Louisiana**, Massachusetts, Montana, New Jersey, **North Dakota, Ohio, Oklahoma, Oregon**, Rhode Island, **Tennessee, Texas, Utah**, West Virginia, **Wisconsin, Wyoming**	Montana

NOTES: The listed jurisdictions include, within their sexual contact and exposure crimes, the requirement that the prohibited activity was done for the purpose of sexual arousal, gratification, degradation or humiliation of the victim or offender. The jurisdictions that are in boldface type under the crime of indecent exposure require that the exposure be done in a public place to be punishable.

*Federal law refers to 18 USC 2241-2247 that applies in special maritime and territorial jurisdictions of the United States or in a federal prison.

SOURCE: Tracy et al. (2012, pp. 29-30).

Use of Force

Jurisdictions have differing characterizations of what constitutes force and what type of force is required to make penetration a crime. "Statutory definitions of force include physical force, violence, force required to overcome victim resistance, or stated or implied threats that place an individual in fear of immediate death or (serious) physical injury to the individual or to a third party, or retaliation" (Tracy et al., 2012, p. 18). Courts have to interpret the relevant statutes and determine if the evidence substantiates

TABLE 2-2 Sufficiency Requirements on the Use of Force for Criminalizing Rape and Sexual Assault, by Jurisdiction

Force	Jurisdiction
Actual force against victim	All
Threatened force against victim is sufficient	All
Force or threat of force against a third party is sufficient	All except Arizona, Arkansas, Georgia, Indiana, Kansas, Louisiana, Massachusetts, Mississippi, Missouri, New Jersey, Ohio, Oklahoma, Tennessee, Vermont, Wisconsin, Puerto Rico, Virgin Islands
Injury required as part of the forcible offenses	District of Columbia, Iowa, New Mexico, North Carolina, Vermont, Washington, West Virginia, Wisconsin, Virgin Islands, Uniform Code of Military Justice

SOURCE: Tracy et al. (2012, p. 20).

the "force" requirement in that jurisdiction. Tracy et al. (2012) indicate that jurisdictions have been moving toward a "more expansive definition" that goes beyond physical force.

Table 2-2 provides a summary of the force requirement across jurisdictions. As shown in the table, all jurisdictions include provisions that criminalize penetration if it is accompanied by force or the threat of force to the victim, although nine jurisdictions and the military require that the force actually inflict physical injury. All but 17 jurisdictions also consider force (or threat of force) against a third party as sufficient.

Absence of Consent

The victim's consent or nonconsent to the penetration is another critical element in determining whether that penetration is criminal (see Box 2-1). In defining what *expressing consent* means, statutes include such factors as conveying permission, positive cooperation in the act, an attitude that expresses that cooperation, or with knowledge of the nature of the act. Some jurisdictions have additional requirements regarding the offender's knowledge—that he or she knew or had reason to know that the victim did not consent. In contrast, however, a number of statutes specifically state that a prior social relationship between the victim and the offender does not constitute consent. If consent is obtained through fraud, then some statutes still consider it consent; others do not. A minority of statutes requires words or overt actions to indicate consent.

BOX 2-1
Factors in Determining Consent for Criminalizing Rape and Sexual Assault

A victim may express consent by

- directly conveying permission (affirmative consent required by the District of Columbia, Minnesota, New Jersey, Washington, Wisconsin),
- positive cooperation in the act, or
- an attitude pursuant to an exercise of free will and with knowledge of the nature of the act.

Factors that may constitute nonconsent:

- The victim felt coerced to consent.
- The offender knew or had reason to know that the victim did not consent (some jurisdictions require this knowledge for nonconsent).
- The consent was by fraud. Some jurisdictions consider obtaining consent by fraud as invalid; other jurisdictions still consider that consent valid.

Other factors that some jurisdictions specifically state shal not constitute consent:

- the prior social relationship between offender and victim or
- the victim's manner of dress.

SOURCE: Tracy et al. (2012, pp. 20-22).

Capacity to Consent

Another component of criminality is the capacity of the victim to give consent for penetration. The age of both the victim and offender affect this determination in most jurisdictions. Table 2-3 provides the minimum age of consent of the victim by jurisdiction. The youngest such age is 10, in Georgia. The oldest age is 18, in both Oregon and the Northern Mariana Islands. Most ages fall between 13 and 16, with 13 years as the most common age. Many states also look at the age difference between the underage victim and the offender in assessing criminal liability.

Most statutes provide special protection regarding "consent" for individuals with mental disabilities, but it does not mean that the court will automatically determine that such an individual does not have the capabil-

TABLE 2-3 Minimum Age of Consent for Sexual Penetration, by Jurisdiction

Minimum Age of Consent	Jurisdiction
10	**Georgia**
12	Alabama, Florida, Iowa, Maine, Missouri, Nebraska, **Wisconsin, Uniform Code of Military Justice**
13	Connecticut, Illinois, Maryland, Michigan, Minnesota, New Hampshire, New Jersey, New Mexico, North Carolina, Ohio, Pennsylvania, South Dakota, Tennessee, Virginia, Wisconsin, Virgin Islands
14	Arizona, California, Hawaii, Kansas, **Massachusetts,** Mississippi, Nevada, **Texas,** Utah, **American Samoa,** Guam, Puerto Rico
15	Colorado, Louisiana, North Dakota, Vermont
16	Arkansas, Delaware, District of Columbia, **Georgia,** Idaho, **Massachusetts,** Montana, Oklahoma, Rhode Island, West Virginia, **Wisconsin, American Samoa, Uniform Code of Military Justice**
17	New York, **Texas**
18	Oregon, Northern Mariana Islands

NOTES: "Minimum age of consent" means that younger individuals are deemed not to have the capacity to consent. Boldface type indicates that there are multiple ages of consent within the state or jurisdiction.
SOURCE: Tracy et al. (2012, p. 20).

ity to give consent. Similarly, several jurisdictions define as an "aggregating factor" the sexual assault of a victim of advanced age.[2]

A broad and sometimes controversial reason for a victim's inability to consent includes conditions, such as physical disability, physical incapacity, and unconsciousness. These conditions include victims who are impaired or unconscious as a result of intoxication. As explained by Tracy et al. (2012, p. 26):

> [In] all but two jurisdictions rape and sexual assault statutes criminalize nonforcible rape and sexual assault of victims who are intoxicated. These intoxication statutes address drug and alcohol-facilitated rape and sexual assault in two ways: either by focusing on the cause (i.e., intoxication) of a victim's inability to consent or by focusing on the effects of a victim's inability to appraise the circumstances of an incident, i.e., inability to consent, regardless of the cause. In addition, some jurisdictions specify criminal conduct based on the manner in which the victim became intoxicated. A victim's intoxication may be voluntary (i.e., an offender takes advantage of a victim's pre-existing intoxication) or involuntary (i.e., an offender surreptitiously or forcefully causes the victim's intoxication).

[2]No state has a specific age identified at which a senior no longer has the capacity to consent.

TABLE 2-4 Intoxication and the Incapacity to Consent for Criminalizing Rape and Sexual Assault, by Jurisdiction

Factors Leading to Incapacity to Consent	Jurisdiction
Victim is unconscious.	All jurisdictions
Victim is voluntarily intoxicated.	10 jurisdictions: Arizona, California, Idaho, Kansas, Louisiana, Montana, South Carolina, Washington, Wisconsin, Virgin Islands

NOTE: Georgia has long-standing case law that covers intoxication.
SOURCE: Tracy et al. (2012, pp. 24-27).

Only 10 jurisdictions have statutes that cover victims who are voluntarily intoxicated, while 40 jurisdictions require that a victim was involuntarily intoxicated. However, 38 of those 40 states have other statutes with such language as "inability to appraise" or "inability to control conduct," which can be used without specifically addressing intoxication. Some jurisdictions additionally have statutes that also require that the offender knows that the victim is unable to consent due to intoxication (see Table 2-4).

A final category related to a victim's incapacity to consent is that the offender is in a "position of authority" over the victim. This relationship may be due to blood (incest) or other duty relationships, such as teacher/student; correctional officer/inmate; medical professional/patient; employer/employee.

For Purpose of Arousal or Degradation

There are 14 jurisdictions (Alabama, Arkansas, California, Colorado, District of Columbia, Idaho, Maryland, Montana, Nevada, Utah, West Virginia, Wyoming, and federal and the military) that require that a crime of penetration be perpetrated for the purpose of arousal of the offender (thus excluding such activities as medical examinations). Five of these jurisdictions (District of Columbia, Montana, West Virginia, and federal and the military) allow that degradation of the victim can be a substitute purpose for arousal.

Combining the Elements to Determine the Severity of the Crime

Crimes that do not include penetration (touching, groping, exposing) are classified as misdemeanor offenses in every jurisdiction. The seriousness of crimes that include penetration varies by jurisdiction (see Table 2-5). Penile penetration with force into the vagina, anus, or mouth is a serious crime in every jurisdiction. In 10 jurisdictions (California, Connecticut, Delaware,

TABLE 2-5 Crimes of Penetration with Degrees of Severity

Type of Forced Penetration	Jurisdiction	Offense Grade
Penile/vaginal	All	Highest level
Penile/anal	All	Highest level except Kansas
Penile/oral	All	Highest level except Oklahoma
Object	All except Louisiana and American Samoa	Not highest level—Class B-D felonies and 2nd-4th degree): California, Connecticut, Delaware, Georgia, Missouri, New York, North Carolina, Oregon, American Samoa, Puerto Rico
Other body part (such as finger or fist)	All except Idaho, Kentucky, Louisiana, Maine, Wisconsin, American Samoa, Northern Mariana Islands	Highest level except Alabama, Georgia, Maryland, New York, Pennsylvania (unless victim is a child), Texas

SOURCE: Tracy et al. (2012, p. 31).

Georgia, Missouri, New York, North Carolina, Oregon, American Samoa, and Puerto Rico), penetration by an object is criminalized, but the classification of the crime is of a lesser severity than penile penetration. Eight other jurisdictions consider penetration with an object as a different but equally serious crime as penile penetration. Only two jurisdictions—Louisiana and the Northern Mariana Islands—do not criminalize object penetration.

Penetration by a body part (not a penis), such as a finger, is generally viewed as less serious than penile penetration. Seven jurisdictions (Idaho, Kentucky, Louisiana, Maine, Wisconsin, American Samoa, and the Northern Mariana Islands) do not criminalize penetration by nonpenile body parts. In 15 jurisdictions, a body part (not a penis) is considered a foreign object and subject to statutes for such objects.

Several other issues may affect the criminalization of certain sexual assaults. Rape of a married partner by a spouse was not legally recognized until the 1970s. By 1993, rape and sexual assault of one's spouse was criminal to some degree in all jurisdictions, but a marital relationship may still affect the severity of the crime in some jurisdictions.

Some jurisdictions have specific statutes regarding "gang rape." These statutes may identify these acts as ones of criminal conspiracy and thus connect them to other statutes covering accomplice liability.

CONCLUSION

The statutes that make up current laws on rape and sexual assault—what those terms mean, whether they are criminal offenses, and the seriousness of the crime—are built on origins that conceived of these as offenses against patriarchal inheritance rights and a female's reproductive capacity. These statutes have changed significantly since the 1970s but have changed at different times and in somewhat different ways in different jurisdictions. The language and concepts are confusing, and in trying to understand survey results, it is critical to keep in mind that victims cannot be expected to respond "accurately" to questions using that language.

In reviewing the state statutes on rape and sexual assault, the panel identified a number of commonalities that would be important to include in uniform definitions of rape and sexual assault for a national survey:

- The victimization is not restricted by gender: both a male or female can be victimized, and the offender can be either male or female.
- "Rape" involves a broad range of penetrations, including penetration of the vagina, anus, or mouth, and with a penis, tongue, fingers, or another object.
- The purpose is for sexual arousal or degradation.
- The offender uses force or threat of force, against either the victim or another person.
- The victim does not consent to the sexual activity or does not have the capacity to consent.
- "Sexual assault" includes a fairly wide range of victimizations that involve unwanted nonpenetration sexual contact.

The panel uses the information presented in this chapter to assess the current definition used by BJS in the NCVS and recommend an expanded definition (see Recommendation 10-7). The panel also uses the components listed above to help assess the current wording of survey questions and to devise improved wording about potential victimizations. Tracy et al. (2012, p. 35) stress the importance of wording questions for victims in ways that will allow the victims to better reveal their experiences, which in turn can help improve the justice systems' responses to those crimes:

> Although some jurisdictions' laws have evolved to incorporate our ever-expanding knowledge of rape and sexual assault and offender behaviors, in other jurisdictions, the laws remain sadly outdated in either language or content. The disconnect between the law and reality can play a crucial role in individual victims' perception of whether or not they were victims of a crime and whether they believe they will receive some measure of justice in the legal system. As a result, the ability to develop questions

that will most accurately and successfully reveal a victim's experience will be invaluable to understanding the incidence and prevalence of rape and sexual assault. It will also play an important role in helping allied criminal justice professionals improve their understanding of rape and sexual assault, their responses to reports of such crimes, and their ability to stop serial predators.

3

Data from Law Enforcement Agencies

This chapter describes the Uniform Crime Reports (UCR) Program that is maintained by the Federal Bureau of Investigation (FBI) using data from local police reports. The program provides important data on crime rates in general and on rates of rape and attempted rape in particular. The UCR is composed of two components: the Summary Reporting System, which was established in 1929 (Wolfgang, 1963), and the newer National Incident-Based Reporting System (NIBRS), which is not yet nationally representative. As background to our discussion of the strengths and weaknesses of both systems and the UCR's statistics on rape, we begin with a discussion of local police reports that provide the detailed data that go into the UCR.

POLICE REPORTS

Basic Reports

Local law enforcement officers compile basic information related to incidents of crimes, arrests, and investigations in a police incident report, referred to simply as the police report. Each jurisdiction and law enforcement organization uses its own format for its police report and has varying instructions on what is to be collected, and when and how it should be completed. A police report may include the following types of information:

- the officer's identifier;
- name and contact information for the victim, witnesses, and any other individuals known to be involved;
- basic factual information, including location, date, and timing of the incident;
- narrative(s) of what happened, including statements from the victim and any witnesses;
- anything that the officer(s) observed first-hand; and
- a description of any suspects.

A police report may also include an indicator of case closure, such as an arrest, or that the officer judged the complaint to be unfounded.

Not every "call for service" that an officer attends will result in an official police incident report. In addition to the specific reporting guidelines from the law enforcement agency, individual officers have some discretion in deciding whether to fill out a report. Those officers also have some discretion in classifying the type of crime and deciding that a particular complaint is unfounded.

Underreporting of Rape and Sexual Assault

There is ample evidence that the crimes of rape and sexual assault are substantially undercounted through police reports, and this section details a selection of that evidence. Some undercounts appear to be due to victims' failing to report these crimes to the police, and some to the way in which law enforcement in certain jurisdictions handles both the victims and the police reports of those crimes.

Victims' Failures to Report

Conducted in 1989-1991, the National Women's Study (NWS; see Chapter 5) collected information on rape and sexual assault. It estimated that 84 percent of rape victims did not report their victimization to police (Kilpatrick, Edmunds, and Seymour, 1992). Tjaden and Thoennes (2006) reported a similar percentage (81 percent) of nonreporting from the National Violence Against Women Survey (see Chapter 5).

Other evidence of nonreporting comes from the National Crime Victimization Survey (NCVS; see Chapter 4), which includes information on whether a victim reported the incident to police. A recent special report, *Victimizations Not Reported to the Police, 2006-2010* (Bureau of Justice Statistics, 2012b), provides basic information on the rates of unreported crimes and the reasons the victims did not involve the police (see Table 3-1).

During the 5 years covered in the special report, the Bureau of Justice

TABLE 3-1 Victimizations Not Reported and Reasons, by Type of Crime, 2006-2010 (in percentage)

Type of Crime	Not Reported	Reason Not Reported: Dealt with in Another Way/ Personal Matter	Not Important Enough to Report	Police Would Not or Could Not Help	Fear of Reprisal or Getting Offender in Trouble	Other Reason or No Main Reason
All crimes	58	20	27	31	5	17
Rape and sexual assault	65	20	6	13	28	33
Robbery	41	20	13	34	10	23
Aggravated assault	44	31	16	17	22	15
Personal larceny	41	17	24	43	2	14
Burglary	45	12	27	40	4	17
Motor vehicle theft	17	16	26	30	7	21
Theft	67	16	31	35	3	16

SOURCE: Data from Bureau of Justice Statistics (2012b, p. 4): the data are generated from the National Crime Victimization Survey.

Statistics (BJS) estimated that 65 percent of rapes and sexual assaults had not been reported to police. As shown in the table, when asked why they did not report to the police, an estimated 41 percent of the victims thought that either the police would not or could not help or they had a fear of reprisal or getting the offender in trouble; 20 percent had dealt with it in another way or felt it was a personal matter; and 33 percent had another reason or not one most important reason. Only 6 percent thought that the crime was not important enough to report. The Bureau of Justice Statistics (2012b, p. 2) further reported "from 2005 to 2010, the percentage of victimizations that went unreported due to the belief that the *police would not or could not help* increased from 7% to 20%." In addition, it is important to note that all of these figures exclude rapes and sexual assaults that were not reported to police and also not disclosed in the NCVS.

Wolitzky-Taylor et al. (2011, pp. 809-810) summarized the research on the victim and offender characteristics that were related to the decision not to report rape and sexual violence to police:

> Stranger rapes are more likely to be reported than rapes by acquaintances (Estrich, 1987). Victims who sustain injuries are also more likely to report their rapes (Bachman, 1993, 1998; Lizotte and Wolfson, 1981), as are those whose assailants used weapons during the rape (Amir, 1971; Bachman, 1998; Lizotte and Wolfson, 1981). Furthermore, victim use of alcohol and/or drugs at the time of the assault has been found to be associated with lower likelihood of reporting (Clay-Warner and Burt, 2005; Fisher et al., 2003).
>
> Demographic variables have also been associated with reporting. One study found that married rape victims and highly educated rape victims were more likely to report than unmarried victims and less educated victims (Lizotte, 1985). Another study found that reports were more likely to be made if the perpetrator was African American (Greenberg and Ruback, 1992).

Police Responses to Reports

In September 2010, the Subcommittee on Crime and Drugs of the Senate Committee on the Judiciary held a hearing titled *Rape in the United States: The Chronic Failure to Report and Investigate Rape Cases.*[1] As summarized by one of the witnesses in prepared testimony, witnesses were asked to comment on three factors related to police handling of reported rape crimes that might contribute to an undercount in statistics from law enforcement agencies: police not accepting rape and other sex crimes for

[1]The hearing occurred on September 14, 2010, and its report was published in 2011 as S.Hrg. 111-891.

investigation; police misclassifying rape and other sex crimes as non-crimes; and police "unfounding"[2] rape cases at an extremely high rate (Dempsey, 2010, p. 1).

A very well-publicized example that involved the Philadelphia Police Department was uncovered and reported by the *Philadelphia Inquirer* in the late 1990s. It revealed a 20-year long practice of downgrading rape cases to a noncriminal category. "The methods and motives varied but the result was almost always the same—to shift offenses out of the 'Part I' group of major crimes tallied nationally by the FBI [in the UCR] and watched closely by the media, the public, politicians and the headquarters brass" (Matza, McCoy, and Fazlollah, 1998). The current Philadelphia Police Commissioner testified at the 2010 Senate hearing, saying, "the deliberate downgrading of rape cases in the Philadelphia Police Department in the late 1990s . . . was a pervasive and systemic failure" (Ramsey, 2010, p. 2).

Similar concerns were raised in a 2010 *Baltimore Sun* article. Reviewing 4 years of UCR data, the reporter found that the Baltimore Police Department coded reported rape cases as false or baseless 30 percent of the time, more often than any other city in the country. In addition, the article noted (Fenton, 2010): "[I]n 4 of 10 emergency calls to police involving allegations of rape, officers conclude that there is no need for a further review, so the case never makes it to detectives—a proportion that experts say is disturbingly high."

In Louisiana, a 2009 article in the New Orleans *Time-Picayune* stated (Maggi, 2009): "[M]ore than half the time New Orleans police receive reports of rape or other sexual assaults against women, officers classify the matter as a noncriminal 'complaint.'" UCR statistics for New Orleans for rape and attempted rape showed a sharp decrease from 2007 to 2008, in contrast with data from the Interim Louisiana State University Public Hospital, where rape victims seeking treatment increased during that same period.

Yet another example comes from Detroit. A recent report found that between 9,000 and 11,300 rape kits[3] were stored for many years, untested, by Detroit police (Williams, 2012).

---

[2]Dempsey (2010, pp. 1-2) further elaborates on three possible reasons for the "unfounding": (1) police misconduct, malfeasance, or lack of proper education regarding the investigation of rape; (2) the existence of a "justice gap" indicating the continued need for law reform; and (3) the limited range of categories in the UCR available to police officers in recording case dispositions.

[3]The sexual assault forensic exam kit (commonly referred to as a "rape kit") is the collection of DNA and other forensic evidence, which is then kept by the sexual assault nurse examiner or medical provider until picked up by law enforcement or the crime lab (Rape, Abuse and Incest National Network, n.d.).

UCR REPORTS

The Summary Reporting System

More than 80 years ago, in 1929, the International Association of Chiefs of Police published *Uniform Crime Reporting: A Complete Manual for Police* and instituted an experimental reporting system with the purpose of providing uniform crime statistics across jurisdictions. The FBI began managing that system the following year and published the first bulletin of *Uniform Crime Reports* (Wolfgang, 1963). The FBI has maintained the Summary Reporting System (SRS) since then.[4] (For a history of changes to the SRS system across those seven decades, see Barnett-Ryan, 2007.)

Although reporting is voluntary, approximately 18,000 law enforcement organizations across the United States contribute to the SRS of the UCR (Federal Bureau of Investigation, n.d.-a), based on information taken from their local police incident reports. Ninety percent of all jurisdictions participate in this system, with coverage of urban areas slightly higher than coverage of rural areas. The FBI produces simple aggregations of these monthly reports from jurisdictions, and uses this compiled information to issue several annual statistical publications, with *Crime in the United States* most relevant to this project.

Because crime statutes related to rape and sexual assault vary across jurisdiction, the FBI attempts to provide uniformity in what is reported by these jurisdictions and how incidents are classified in that reporting through a guidance handbook, the *Uniform Crime Reporting Handbook* (Federal Bureau of Investigation, 2004). The most recent edition of the *Handbook* is dated 2004.[5] On its website, the FBI (n.d.-a) discusses the *Handbook*, stating:

> [It] explains how to classify and score offenses and provides uniform crime offense definitions. Acknowledging that offense definitions may vary from state to state, the FBI cautions agencies to report offenses not according to local or state statutes but according to those guidelines provided in the handbook. Most agencies make a good faith effort to comply with established guidelines.

However, because police investigate crimes and complete reports in the framework of their own jurisdictional statutes, law enforcement agencies are being asked to report a summary of these crimes on the UCR based on a different set of definitions characterizing those events. To assess this process

[4]The SRS, unlike the NIBRS, contains only summary-level data, not incident-level data.

[5]In June 2013, the FBI released a *Summary Reporting System (SRS) User Manual,* Version 1.0, as a replacement for the *Handbook*. However, this chapter (and this report) use the *2004 Handbook* as reference because it was the standard during the panel's data gathering period.

the FBI offers a voluntary UCR Quality Assurance Review (QAR) as part of a triennial audit of states' criminal justice information systems (Federal Bureau of Investigation, 2004, p. 3):

> The purpose of the QAR is to ensure that each state UCR Program adheres to summary and incident-based reporting methods that are consistent with UCR standards in order to achieve uniform crime reporting nationwide. In 2001, the QAR incorporated a statistical sampling methodology to select records for data quality review and to project the number of discrepant crime reports a state UCR Program submits to the national UCR Program.

The panel did not find any public information that includes or summarizes results from these quality reviews.

The guidelines in the *Handbook* divide crimes into two categories: major crimes (known as Part I crimes) to be recorded as "crimes known to police" and other crimes (Part II) ranging from minor assault to vagrancy.[6] "The Part I offenses came to be used as a crime index, much like a price or cost-of-living index" (Wolfgang, 1963, p. 709).[7] Part II offenses are reported on the UCR only after arrests are made.

Law enforcement agencies submit a monthly summary of crimes for the SRS component of the UCR. The primary reporting form,[8] *Return A—Monthly Return of Offenses Known to the Police*, lists seven major (Part I) crime categories: criminal homicide, forcible rape, robbery, assault, burglary, larceny, and motor vehicle theft. Each major category has several subcategories (see Figure 3-1). For each major category and subcategory, the form requests the following information:

- number of offenses reported or known to police (includes "unfounded" and attempts);
- number of unfounded offenses (i.e., false or baseless complaints);
- number of actual offenses (unfounded offenses subtracted from the total);
- number of offenses cleared by arrest or exceptional means; and
- number of clearances involving only persons under 18 years of age.

---

[6]Part II offenses include other assaults; forgery and counterfeiting; fraud; embezzlement; stolen property—buying, receiving, possessing; vandalism; weapons—carrying, possessing, etc.; prostitution and commercialized vice; sex offenses; drug abuse violations; gambling; offenses against the family and children; driving under the influence; liquor laws; drunkenness; disorderly conduct; vagrancy; all other offenses; suspicion; curfew and loitering laws; and runaways.

[7]The systematic downgrading of reports of rape from a Part I crime to a Part II crime in certain jurisdictions was the subject of newspaper investigations described above.

[8]Several other forms are submitted monthly for the SRS, but they are not relevant to the crimes of rape and sexual assault.

RETURN A - MONTHLY RETURN OF OFFENSES KNOWN TO THE POLICE

1-720 (Rev. 02-22-13)
OMB No. 1110-0001
Expires 07-31-16

This report is authorized by law Title 28, Section 534, U.S. Code. Your cooperation in completing this form will assist the FBI, in compiling timely, comprehensive, and accurate data. Please submit this form monthly, by the seventh day after the close of the month, and any questions to the FBI, Criminal Justice Information Services Division, Attention: Uniform Crime Reports/Module E-3, 1000 Custer Hollow Road, Clarksburg, West Virginia 26306; telephone 304-625-4830, facsimile 304-625-3566. Under the Paperwork Reduction Act, you are not required to complete this form unless it contains a valid OMB control number. The form takes approximately 10 minutes to complete. Instructions for preparing the form appear on the reverse side.

CLASSIFICATION OF OFFENSES	DATA ENTRY	2 OFFENSES REPORTED OR KNOWN TO POLICE (INCLUDE "UNFOUNDED" AND ATTEMPTS)	3 UNFOUNDED, I.E., FALSE OR BASELESS COMPLAINTS	4 NUMBER OF ACTUAL OFFENSES (COLUMN 2 MINUS COLUMN 3) (INCLUDE ATTEMPTS)	5 TOTAL OFFENSES CLEARED BY ARREST OR EXCEPTIONAL MEANS (INCLUDES COL. 6)	6 NUMBER OF CLEARANCES INVOLVING ONLY PERSONS UNDER 18 YEARS OF AGE
1. CRIMINAL HOMICIDE a. MURDER AND NONNEGLIGENT HOMICIDE (Score attempts as aggravated assault) If homicide reported, submit Supplementary Homicide Report	11					
b. MANSLAUGHTER BY NEGLIGENCE	12					
2. RAPE TOTAL	20					
a. Rape	21					
b. Attempts to Commit Rape	22					
Historical Rape (See Instruction #15 below)						
3. ROBBERY TOTAL	30					
a. Firearm	31					
b. Knife or Cutting Instrument	32					
c. Other Dangerous Weapon	33					
d. Strong-Arm (Hands, Fists, Feet, Etc.)	34					
4. ASSAULT TOTAL	40					
a. Firearm	41					
b. Knife or Cutting Instrument	42					
c. Other Dangerous Weapon	43					
d. Hands, Fists, Feet, Etc. - Aggravated injury	44					
e. Other Assaults - Simple, Not Aggravated	45					
5. BURGLARY TOTAL	50					
a. Forcible Entry	51					
b. Unlawful Entry - No Force	52					
c. Attempted Forcible Entry	53					
6. LARCENY - THEFT TOTAL (Except Motor Vehicle Theft)	60					
7. MOTOR VEHICLE THEFT TOTAL	70					
a. Autos	71					
b. Trucks and Buses	72					
c. Other Vehicles	73					
GRAND TOTAL	77					

CHECKING ANY OF THE APPROPRIATE BLOCKS BELOW WILL ELIMINATE YOUR NEED TO SUBMIT REPORTS WHEN THE VALUES ARE ZERO. THIS WILL ALSO AID THE NATIONAL PROGRAM IN ITS QUALITY CONTROL EFFORTS.

- ☐ NO SUPPLEMENTARY HOMICIDE REPORT SUBMITTED SINCE NO MURDERS, JUSTIFIABLE HOMICIDES, OR MANSLAUGHTERS BY NEGLIGENCE OCCURRED IN THIS JURISDICTION DURING THE MONTH.
- ☐ NO SUPPLEMENT TO RETURN A REPORT SINCE NO CRIME OFFENSES OR RECOVERY OF PROPERTY REPORTED DURING THE MONTH.
- ☐ NO LAW ENFORCEMENT OFFICERS KILLED OR ASSAULTED REPORT SINCE NONE OF THE OFFICERS WERE ASSAULTED OR KILLED DURING THE MONTH.
- ☐ NO AGE, SEX, AND RACE OF PERSONS ARRESTED UNDER 18 YEARS OF AGE REPORT SINCE NO ARRESTS OF PERSONS WITHIN THIS AGE GROUP.
- ☐ NO AGE, SEX, AND RACE OF PERSONS ARRESTED 18 YEARS OF AGE AND OVER REPORT SINCE NO ARREST OF PERSONS WITHIN THIS AGE GROUP.
- ☐ NO MONTHLY RETURN OF ARSON OFFENSES KNOWN TO LAW ENFORCEMENT REPORT SINCE NO ARSONS OCCURRED.

DO NOT USE THIS SPACE	
	INITIALS
RECORDED	
EDITED	
ENTERED	
ADJUSTED	
CORRES	

Month and Year of Report

Agency Identifier

Population

Prepared by

Title

Telephone Number

Date

Agency and State

Chief, Sheriff, Superintendent, or Commanding Officer

FIGURE 3-1 Uniform Crime Reporting System, *Return A*, Monthly Reporting Form (pp. 1-2).

NOTE: This form was revised on February 22, 2013, following the new definition of rape. A line for "Historical Rape" and Instruction #15 were added.

SOURCE: Federal Bureau of Investigation (n.d.-b).

INSTRUCTIONS FOR PREPARING *RETURN A*

(Detailed instructions are given in the *Uniform Crime Reporting Handbook*)

1. All Offenses listed on the *Return A* which occur during the month should be scored whether they become known to the police as the result of:

   a. Citizens' complaints.
   b. Reports of police officers.
   c. "On view" (pick-up) arrests.
   d. Citizens' complaints to sheriff, prosecutor, county police, private detectives, constables, etc.
   e. Any other means.

2. The offenses listed in Column 1 are the Part I offenses of the Uniform Crime Reporting Program plus the offenses of simple assault and manslaughter by negligence. Follow the instructions for classifying and scoring as presented in the *Uniform Crime Reporting Handbook*. Offenses committed by juveniles should be classified in the same manner as those committed by adults even though the juveniles may be handled by juvenile authorities.

3. Adjustments should be made on this month's return for offenses omitted or scored inaccurately on returns of preceding months or those now determined to be unfounded. Offenses that occurred in a previous month but only became known to you this month should be scored this month.

4. Consider all spaces for each classification of offenses in Columns 2, 3, 4, 5, and 6. The breakdowns for rape, robbery, assault, burglary, and motor vehicle theft, when added should equal the total for each of these offenses. Do not enter zeroes where no count exists.

5. Attempts of rape, robbery, assault, burglary, larceny-theft, and motor vehicle theft are to be scored on this form.

6. **Column 2:** Enter opposite the proper offense classification the total number of such offenses reported or known through any means. "Unfounded" complaints are included. Attempts are included except in homicide classifications.

7. **Column 3:** Enter the number of complaints which were proven to be "unfounded' by police investigation. An "unfounded" offense is one in which a complaint was received, but upon investigation, proves either to be baseless or not to have actually occurred. Remember that recovery of property or clearance of an offense does not unfound a complaint.

8. **Column 4:** Number of actual offenses. This number is obtained by subtracting the number in Column 3 from that in Column 2.

9. **Column 5:** Enter the total number of offenses **cleared** during the month. This total includes the clearances which you record in Column 6. An offense is cleared when one or more persons are charged and turned over for prosecution for that offense. Clearance totals also include exceptional clearances which are explained in the *Uniform Crime Reporting Handbook*.

10. **Column 6:** Enter here the number of offenses which are cleared through the arrest, releasing to parents, or other handling of persons under the age of 18. In those situations where an offense is cleared through the involvement of both an adult and a person under 18 years of age, count the clearance only in Column 5.

11. The grand totals for Columns 2, 3, 4, 5, and 6 are the totals of each of the seven classifications.

12. Tally books can be used to maintain a running count of offenses through the month. Totals for the *Return A* can then be taken directly from the Tally book. These Tally books can be obtained by corresponding with the FBI, Criminal Justice Information Services Division, Attention: Uniform Crime Reports/Module E-3, 1000 Custer Hollow Road, Clarksburg, West Virginia 26306; telephone 304-625-4830, facsimile 304-625-3566.

13. This *Return A* report should be forwarded to the FBI Uniform Crime Reports even though no offenses of this type listed were committed during the month. However, it is not necessary to submit supplemental reports in such cases. Simply check the appropriate box within the block near the bottom of the *Return A* report.

14. Any inquiry regarding the completion of this form, the classification and scoring of offenses, or prior to submitting crime data by computer printout, contact the Uniform Crime Reporting Program at the above-mentioned address.

15. Reporting according to the historical definition of rape is optional. This count should be included in the total reported in line 2a or line 2b. It is intended to be used for trending purposes.

It is important to note that these are the only crimes that are reported to the FBI on the *Return A*, and thus the only crimes (except arson)[9] that are included in the crime rates published in *Crime in the United States*.

When multiple Part I crimes are committed simultaneously by the same

[9]Arson is a Part I crime, but data on it are collected on a separate form instead of the *Return A*.

offender(s) on the same victim in the same incident, the *Handbook* provides a hierarchy rule and states that the "multiple crime" should be categorized as the single crime highest in the hierarchy list.[10] This rule ensures that the incident is reported only once on the *Return A*. For example, a rape/homicide would be classified as a homicide, and a rape/robbery would be classified as a rape.

Definition of Rape in the UCR SRS

In regard to rape and sexual assault, it is important to consider the definitions used in the UCR. Under forcible rape, there are two subcategories: "rape by force" and "attempts to commit forcible rape." The FBI provides a definition of forcible rape in the 2004 version of the *Handbook* (the latest as of the development of this report) as the "carnal knowledge of a female forcibly and against her will" (Federal Bureau of Investigation, 2004, p. 19). The *Handbook* further explains that "carnal knowledge" is penetration (however slight) of a penis into a vagina. This definition of rape in the UCR has remained essentially unchanged since 1929, and the category of sexual assault does not exist in the UCR. Thus, sexual assaults that do not meet the above narrow definition of rape are included with the more general category of "aggravated assault" as a Part I crime or as "other assaults" as a Part II crime.

Discussions to update the definition of forcible rape used by the UCR began in 1999 and moved through a series of reviews, input from police organizations, further discussions, and public hearings (Federal Bureau of Investigation, 2013); a revision was announced on January 6, 2012, by Attorney General Eric Holder. The revision, which is designed to lead to a more comprehensive reporting of rape through the UCR, is more aligned with changes that individual jurisdictions have been making in criminal statutes since the 1970s. The new definition of forcible rape is as follows (Federal Bureau of Investigation, 2012):

> The penetration, no matter how slight, of the vagina or anus with any body part or object, or oral penetration by a sex organ of another person, without the consent of the victim.

The new definition is a substantial change from the older definition: it recognizes that males as well as females may be rape victims; it includes penetration in the anus and mouth as well as in the vagina; and it in-

[10]Hierarchy List of Part I crimes, in descending order: criminal homicide, forcible rape, robbery, aggravated assault, burglary, larceny-theft (except motor vehicles), motor vehicle theft, and arson. The *Handbook* hierarchy list also includes subcategories within each of these major categories.

cludes penetration by body parts other than a penis or by objects. This new definition was expected to be used operationally on the *Return A* beginning January 2013.

Statistics from the UCR SRS

The FBI annually publishes *Crime in the United States* based on data from the UCR SRS. The most recent edition is for 2011, and therefore the statistics in this volume reflect the pre-2012 definition of forcible rape. That is, these counts include forcible rape and attempted rape as described above. In 2011, the UCR counted 83,425 forcible rapes (see Table 3-2 and Figure 3-2 for data from 1992 through 2011). Because of the narrow definition of rape and the likely underreporting of these victimizations (discussed above),

TABLE 3-2 Counts of Forcible Rapes and Attempted Forcible Rapes, Uniform Crime Reports Summary Reporting System

Year	Population	Forcible Rapes and Attempted Rapes	Rate per 1,000 People
1992	255,029,699	109,062	0.428
1993	257,782,608	106,014	0.411
1994	260,327,021	102,216	0.393
1995	262,803,276	97,470	0.371
1996	265,228,572	96,252	0.363
1997	267,783,607	96,153	0.359
1998	270,248,003	93,144	0.345
1999	272,690,813	89,411	0.328
2000	281,421,906	90,178	0.320
2001	285,317,559	90,863	0.318
2002	287,973,924	95,235	0.331
2003	290,788,976	93,883	0.323
2004	293,656,842	95,089	0.324
2005	296,507,061	94,347	0.318
2006	299,398,484	94,472	0.316
2007	301,621,157	92,160	0.306
2008	304,059,724	90,750	0.298
2009	307,006,550	89,241	0.291
2010	309,330,219	85,593	0.277
2011	311,591,917	83,425	0.268

NOTE: Population numbers are U.S. Census Bureau provisional estimates as of July 1 for each year except 2000 and 2010, which are decennial census counts. The rate per 1,000 people is based on those population counts.
SOURCE: Data from online Uniform Crime Reports Data Tool, Federal Bureau of Investigation (n.d.-c).

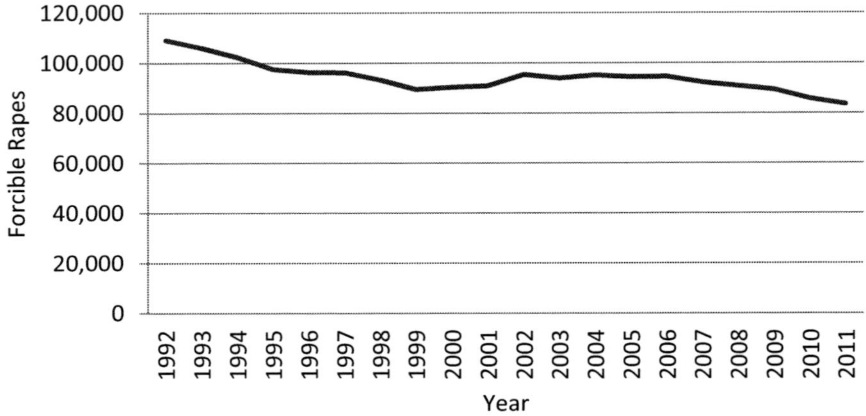

FIGURE 3-2 Forcible rapes and attempted rapes, Uniform Crime Reports Summary Reporting System.
SOURCE: Data from online Uniform Crime Reports Data Tool, Federal Bureau of Investigation (n.d.-c).

the numbers significantly underestimate the number of rape crimes. Other types of sexual assault are not included in these numbers.

The UCR SRS documentation is fairly clear in how these statistics are tabulated based on the reports submitted to the FBI from various law enforcement agencies. However, the details of how these law enforcement agencies actually complete their monthly reports are very murky. It is particularly unclear how an agency would categorize a victimization that is considered rape under state/local statutes but not under the FBI definition. Is this situation handled consistently across agencies? This is an important question that needs more transparency.

The NIBRS Component

The National Incident-Based Reporting System is a modern addition to the UCR system that reflects several innovations in comparison with the SRS. It is incident based (i.e., a record is submitted for each incident), rather than based on a monthly summary of incidents. This approach provides the opportunity to include more important information about each incident that cannot be included in a summary system. A second innovation is that the NIBRS reports on 46 different types of offenses.

Currently, only 6,444 law enforcement agencies representing approximately 25 percent of the U.S. population report through NIBRS instead of the SRS. FBI personnel generate monthly reports for those agencies from NIBRS for the SRS. Because of the relatively few agencies that report through NIBRS, it cannot be used for meaningful national-level crime statistics.[11] Consequently, the panel did not use data from NIBRS in this report. Many users of UCR crime data would like to see this system grow and develop so that it replaces the SRS.

[11]The panel is unaware of any existing statistical procedure for extrapolating the NIBRS data to the entire United States.

4

National Crime Victimization Survey

The National Crime Victimization Survey (NCVS) is a national household survey, conducted on an ongoing basis by the Census Bureau for the Bureau of Justice Statistics (BJS). It is the major tool used by BJS to provide statistics on criminal victimizations, covering rape, sexual assault, robbery, assault, theft, household burglary, and motor vehicle theft. It is the source of data for an annual BJS publication, *Criminal Victimization*, and numerous other regular and special reports.

Over its history, the NCVS (and its predecessor the National Crime Survey) has been a uniquely valuable source of information on the "dark figure of crime"—those crimes not reported to police (Baumer and Lauritsen, 2010; Biderman and Reiss, 1967; Lauritsen and Heimer, 2008; National Research Council, 2008). The NCVS is complementary to, and frequently compared to, data from the Uniform Crime Reports (UCR) discussed in Chapter 3. The NCVS enables BJS to provide statistics on victimization for the U.S. population (12 years of age and older) as a whole and also for important subpopulations, such as women, the elderly, members of various racial groups, city dwellers, and other groups definable from survey data. It also focuses attention on the characteristics of the victims of crime, including the nature of their victimizations.

This chapter first briefly describes the history and development of the NCVS, including a major redesign in 1991. It then provides some descriptive details about the survey, including the target population and sample design, the data collection process, the survey instruments, the definitions used for rape and sexual assault, the estimation process, and the products. This chapter is descriptive in nature: it does not offer any assessment of the quality of the survey. (The panel's assessment of various features of the survey is the subject of Chapters 7, 8, and 9.) In reading this chapter, it is important to keep in mind that the NCVS is an omnibus survey, covering many types of criminal victimizations, not just rape and sexual assault. However, to the extent possible and in keeping with the panel's charge, this chapter and the entire report are focused on the issues relevant to those two victimizations.

HISTORY AND DEVELOPMENT

Origins and Early Development: 1967-1991

The NCVS has its roots in a report from President Johnson's Commission on Law Enforcement and Administration of Justice in 1967. The commission reported that the UCR was useful in many ways, but it noted several critical concerns with having just one source of crime statistics. One concern was that the UCR was a summary of only those crimes reported to (and recorded by) police and not all crimes (see Chapter 3). Secondly, the commission found that the UCR administrative statistics were open to possible manipulation and misrepresentation. Finally, the commission said, the UCR lacked information about the victims, the victimization incidents, and the offenders that is needed to develop effective policy choices. The commission recommended the development of a national crime survey (President's Commission on Law Enforcement and the Administration of Justice, 1967; Rennison and Rand, 2007).

Work on the National Crime Survey (NCS), predecessor of the NCVS, began soon afterwards with small-scale field tests of questions asked of victims of crimes that had been reported to police. A questionnaire was developed from these tests, which was initially implemented as a supplement to the Census Bureau's Quarterly Household Survey in 1971. These supplements served as a way to further develop concepts and questions; they were not used to produce BJS published reports.

In July 1972 the new NCS was fielded. The core had a sample of 72,000 households and noninstitutionalized group quarters.[1] In addition, the suite of surveys included a national sample of 15,000 businesses, as well as city-based samples in each of 26 major cities (12,000 households and 2,000 businesses) to support local-area estimates of victimization. These

[1]Group quarters included group living arrangements, generally of individuals who are not related to each other, such as college residence halls, residential treatment centers, etc.

"cities surveys" began in eight "impact" cities and were expanded to other cities over the next 3 years.[2]

A change in design was recommended a few years later in *Surveying Crime* (National Research Council, 1976). One of the recommended changes was to abandon the "cities surveys" to provide resources to continue to support the core national household sample. BJS established a redesign consortium, and small-scale changes to the survey and its instruments were implemented in 1986.

Considerable cognitive work was undertaken as part of this redesign, much of which focused on understanding and testing screener strategies to increase reported incidents. In the NCS the screener directly asked a respondent about several (but not all) types of possible victimizations. According to the Bureau of Justice Statistics (1989, pp. 11-12):

> The major proposed changes that were ultimately tested [for this omnibus survey] included asking respondents about victimizations that occurred in various life "domains" such as work and leisure, providing many short cues to help trigger memory of incidents in these life contexts, and attempting to evoke the sort of emotional states that might result from a crime incident (such as anger or fear) before administering the screen questions. The aim of these innovations was both to elicit increased reporting of crime incidents and to structure the recall task to a greater degree, so that cognitive and subcultural differences among respondents would have a smaller impact on the reporting of crime incidents.

Other topics, such as bounding, reference period, interview-to-interview recounting, ways to enhance the reliability of dating incidents, and series crimes were analyzed during this redesign process. More details are available from Bureau of Justice Statistics (1989). Following this development, a three-wave national pretest of the redesigned victimization survey was fielded in 1989.

Development Since the 1991 Redesign

In July 1991, the name of the redesigned survey was changed to the National Crime Victimization Survey. The new survey questionnaire included revisions that were viewed as a major improvement over its predecessor (Bachman and Taylor, 1994). Specifically, the redesigned questionnaire asked respondents about experiencing forced or coerced sexual behavior in addition to having been "attacked" or "threatened." The redesign process

---

[2]The impact cities were Atlanta, Baltimore, Cleveland, Dallas, Denver, Newark, Portland (Oregon), and St. Louis. More information about this early version of the survey can be found in Lynch and Addington (2007).

also examined the appropriate reference period to use. Based on cognitive research demonstrating that long recall periods yielded poorer reports and that a reference period of 12 months generally reduced the reporting of all criminal incidents by approximately 30 percent (Cantor and Lynch, 2000), a 6-month reference period was retained.

The NCVS was fully implemented in 1993 with 70,707 households and 136,747 individuals interviewed. Because the cost of conducting the NCVS grew while congressional funding remained essentially flat for BJS, sample size cuts were implemented to the NCVS over the subsequent 15 years: 12 percent in 1996, 4 percent in 2002, 16 percent in 2006, and 14 percent in 2007 (National Research Council, 2008).

Surveying Victims: Options for Conducting the National Crime Victimization Survey (National Research Council, 2008, p. 78) raised questions about the funding levels of the NCVS: "As currently configured and funded, the NCVS is not achieving and cannot achieve BJS's legislatively mandated goal." The report then made two strong recommendations (National Research Council, 2008, p. 79):

1. BJS must ensure that the nation has quality annual estimates of levels and changes in criminal victimization.
2. Congress and the administration should ensure that BJS has a budget that is adequate to field a survey that satisfies that goal.

BJS received increased funding for the NCVS, and sample sizes were raised. In 2011, 79,802 households and 143,122 individuals were interviewed.

TARGET POPULATION AND SAMPLE DESIGN

The target population of the NCVS is the noninstitutionalized population of the United States, 12 years of age or older. It includes (Bureau of Justice Statistics, 2008b, p. 1)

> residents living throughout the United States, including persons living in group quarters, such as dormitories, rooming houses, and religious group dwellings. Crew members of merchant vessels, Armed Forces personnel living in military barracks, and institutionalized persons, such as correctional facility inmates, were not included in the scope of this survey. Similarly U.S. citizens residing abroad and foreign visitors to this country were excluded.

The design challenge for the NCVS is to estimate the number of victimizations (for the target population) that have occurred within the past year and the change from the previous year. The victimizations include events that range from statistically common, such as theft (104.2 per 1,000 households in 2011) to events that are uncommon, such as total violent crime (22.5 per 1,000 people [12+ years] in 2011) to events that are statistically very rare, such as rape and sexual assault (0.9 per 1,000 people [12+ years] in 2011). This type of situation is discussed in a classic textbook on sampling (Hansen, Hurwitz, and Madow, 1953, p. 107): "the particular population under consideration is buried in a very much larger population, and at best it will be an expensive proposition to obtain a sample that will give estimates . . . that will be of a high relative precision."

To produce estimates for a wide range of victimizations, the NCVS uses a classical area sampling design, with selection using multiple stages with stratification and clustering (see detailed description below). The result is a nationally representative sample of individuals identified for data collection.

Table 4-1 provides a summary of the two main stages of this multistage design. The primary sampling units (PSUs) in the first selection stage are small groups of neighboring counties, or large individual counties or metropolitan areas. The largest PSUs (in terms of population size) are included in the NCVS sample with certainty. The remaining PSUs are grouped into strata with similar geographic and demographic characteristics[3] as determined by the most recent census, and then sampled with probability proportional to their population size.

Secondary sampling units (SSUs) are individual residential units, which are sampled within each selected PSU separately from each of four nonoverlapping list frames of eligible units. The four frames (called unit, permit, area, and noninstitutionalized group quarters) are used to maximize the population coverage of the NCVS sample. "Unit" listings are those residential units included in the Master Address File developed for the 2000 census.[4] "Permit" listings represent new construction and are taken from building permits for new residential units since the 2000 census. "Area" listings are generated from periodic canvassing of selected census blocks within the PSUs and consist of identified residential units that are not on the other three frames. Noninstitutionalized "group quarters" are those living quarters identified for the 2000 census "where people live or stay, in a group living arrangement, that is owned or managed by an entity or organization providing housing and/or services for the resident. . . . People

[3]These stratification characteristics include geographic region, population density, rate of growth, population, principal industry, and type of agriculture.

[4]The Master Address File developed for and used in the decennial census is essentially the listing of all known living quarters in the United States, but excluding group quarters. Address listings from the 2000 census, rather than the 1990 census, began to be used in the NCVS sample in 2005.

TABLE 4-1 NCVS Sample Design Summary

Stage	Sampling Unit and Frame Source	Stratification
1	**Primary Sampling Unit (PSU):** Counties, groups of counties, or large metropolitan areas. The largest PSUs are self-representing (SR) PSUs, chosen with certainty into the PSU sample; all other PSUs are non-SR (NSR) and thus sampled with probabilities less than 1. **Frame:** Listing of all counties, groups of counties, or large metropolitan areas in the entire United States. (Military bases and external territories are not included.)	Each SR PSU is considered to be a unique stratum. NSR PSUs are stratified by similar geographic and demographic characteristics.
2	**Secondary Sampling Unit (SSU):** "Measures" defined as clusters of four neighboring housing units from the unit listing in a specific frame. When one "measure" is selected systematically, the next six "measures" in the listing are also selected. **Frame:** Four nonoverlapping lists of the following types were created in each sample PSU: unit, area, permit, and noninstitutional group quarters. "Unit" listings are those residential units included in the Master Address File developed for the 2000 census. "Permit" listings represent new construction and are taken from building permits for new residential construction since the 2000 census. "Area" listings are generated from periodic canvassing of selected census blocks within the PSUs and consist of identified residential units that are not on the other three frames. Noninstitutional "group quarters" are those living quarters identified for the 2000 census "where people live or stay, in a group living arrangement, that is owned or managed by an entity or organization providing housing and/or services for the resident."	**Explicit Stratification:** By type of frame.

NOTES: Target population is the noninstitutionalized U.S. residents aged 12+ years in all 50 states. The noninstitutionalized population includes people living throughout the United States, including persons living in group quarters, such as dormitories, rooming houses, and religious group dwellings. Not included are crew members of merchant vessels; Armed Forces personnel living in military barracks; institutionalized persons, such as correctional facility inmates; U.S. citizens residing abroad; and foreign visitors to the United States.
SOURCE: Information from Bureau of Justice Statistics (2008b).

Sample Selection	Sample Size
SR PSUs selected with probability equal to 1. In each NSR stratum: • 1 PSU is randomly chosen from each stratum with probability proportional to size (PPS). • The measure of size for PPS selection is the PSU's population size.	• Originally, the total PSU sample size was 245. o 93 SR PSUs were identified. o 152 NSR PSUs were randomly selected. • In October 1996, the PSU sample size was reduced to 203. o The same 93 SR PSUs were retained. o aThe number of NSR PSUs was reduced to 110.
• Within a selected PSU, each of the four frames is sampled separately. The housing units in a given frame are grouped together into clusters of four units called "measures." The second stage sample is drawn with a systematic selection of "measures" from each frame. When one "measure" is selected, the sample includes that "measure" and the next six "measures" in the listing. • The household sample is divided into six "rotation groups," and each rotation group is in turn randomly divided into six "panels." • The different panel of households (from each rotation group) is interviewed each month during a 6-month data collection period. • All residents 12+ years at each address for each type of listing are selected for interview.	• Number of participating households varies from year to year but is currently about 80,000 households. • The number of survey respondents varies per year, ranging from 134,000 to 181,000. Currently it is approximately 145,000. • Annual overall response rates have varied between 86%-91%; currently it is approximately 87%.

living in group quarters are usually not related to each other" (U.S. Census Bureau, 2010).[5]

Within a selected PSU, each of the four frames discussed above is sampled separately.[6] The housing units in a given frame are grouped together into clusters of four units. The second stage sample is drawn with a systematic selection of these clusters from each frame.[7] When one cluster is selected, the sample includes that cluster and the next six clusters in the listing. This sampling process is similar to that used for the Current Population Survey.[8] The actual number of households and persons interviewed in the NCVS sample varies slightly from year to year (see Table 4-2).

For field interviewing, the overall sample is divided into six rotation groups, and each rotation group is further divided into six equal collection panels. A single panel from each rotation group is contacted each month, allowing the entire rotation group to be contacted within 6 months with an equal distribution across months. Table 4-3 displays the rotation pattern for a single rotation group, showing the interview month (I) and the months (x) that are included in the reference period for that interview. Because the survey is continuous, newly constructed housing units are selected as described and assigned to rotation groups and panels for subsequent incorporation into the sample. A new rotation group enters the sample every 6 months, replacing a group phased out. With this overall design, each household in a rotation group is interviewed once every 6 months for a total of seven interviews.[9] The initial interview is used to bound information reported in subsequent interviews.[10] Prior to 2007 data from the initial interview were

[5]Noninstitutionalized group quarters include such living quarters as college residence halls, group homes, emergency and transitional shelters, religious group homes, and workers' dormitories (U.S. Census Bureau, 2010). The NCVS sample excludes other specified institutionalized group quarters, such as military barracks, merchant vessels, correctional facilities, halfway houses, and skilled nursing facilities (Bureau of Justice Statistics, 2008b).

[6]It was unclear from the documentation whether these frames were sampled proportionately, or disproportionately based on some criteria. As the report discusses later, a disproportionate sample weighted toward certain group quarters or PSUs in more at risk areas may be appropriate for measuring rape and sexual assault.

[7]Before sampling, the group quarters frame "is converted to housing unit equivalents because Census addresses of individual group quarters or people within a group quarter are not used in the sampling. The number of housing unit equivalents is computed by dividing the Census 2000 group quarters population by the average number of people per household" (U.S. Census Bureau, 2006, pp. 3-8).

[8]More documentation of the process can be found in U.S. Census Bureau (2006).

[9]Respondents are contacted every 6 months, and because the data collection is retrospective, they are asked for a total of 3.5 years of data over the 3 years of contact.

[10]Bounding recall procedures are used to minimize forward telescoping. Forward telescoping describes a pattern of reporting events as having occurred more recently that they actually did. Bounded recall requires a panel survey; in every interview after the first, the interviewer checks whether any incidents the respondent reports were also reported in the previous interview in order to eliminate double reporting.

TABLE 4-2 Sample Size and Response Rates for Households and Persons in the NCVS: 1993-2011

Year	Eligible Households[a]	Household Interviewed	Household Response Rate (%)	Eligible Persons[b]	Persons Interviewed[c]	Response Rate for Persons (%)[d]
1993	74,025	70,707	95.5	146,719	136,747	93.2
1994	99,817	94,978	95.2	196,865	181,205	92.0
1995	100,824	95,504	94.7	197,366	179,816	91.1
1996	97,692	90,779	92.9	188,010	170,655	90.8
1997	90,536	85,821	94.8	177,603	158,939	89.5
1998	91,402	86,309	94.4	177,654	157,797	88.8
1999	91,831	85,789	93.4	175,524	155,501	88.6
2000	92,934	86,800	93.4	177,924	159,420	89.6
2001	93,935	87,360	93.0	179,059	159,900	89.3
2002	91,669	84,685	92.4	174,252	152,105	87.3
2003	91,296	83,659	91.6	172,703	149,040	86.3
2004	92,423	84,361	91.3	173,796	148,577	85.5
2005	85,072	77,224	91.0	158,988	134,041	84.3
2006	83,604	75,979	90.9	157,108	135,264	86.1
2007	91,774	82,905	90.3	170,869	147,296	86.2
2008	84,186	76,128	90.4	155,704	134,179	86.2
2009	84,410	77,455	91.8	157,796	137,329	87.0
2010	88,823	81,948	92.3	167,444	146,567	87.5
2011	88,583	79,802	90.1	162,867	143,122	87.9

[a]The number of eligible households in which there was a household response at least once during that year.
[b]The number of persons in responding households.
[c]The number of eligible household members who responded at least once during the year.
[d]The rate for person response in a responding household.
SOURCE: Data provided to panel by Bureau of Justice Statistics (personal communication).

TABLE 4-3 Panel Design for the NCVS

Interview Month (**I**) and Reference Months for Reporting (x) by Data Collection Panels (P1-P6) Within a Rotation Group

	Months in Year 1												Months in Year 2											
	1	2	3	4	5	6	7	8	9	10	11	12	1	2	3	4	5	6	7	8	9	10	11	12
P1	x	x	x	x	x	I	x	x	x	x	x	I	x	x	x	x	x	I	x	x	x	x	x	I
P2		x	x	x	x	x	I	x	x	x	x	x	I	x	x	x	x	x	I	x	x	x	x	x
P3			x	x	x	x	x	I	x	x	x	x	x	I	x	x	x	x	x	I	x	x	x	x
P4				x	x	x	x	x	I	x	x	x	x	x	I	x	x	x	x	x	I	x	x	x
P5					x	x	x	x	x	I	x	x	x	x	x	I	x	x	x	x	x	I	x	x
P6						x	x	x	x	x	I	x	x	x	x	x	I	x	x	x	x	x	I	x

	Months in Year 3												Months in Year 4											
P1	x	x	x	x	x	I	x	x	x	x	x	I	x	x	x	x	x	I						
P2	I	x	x	x	x	x	I	x	x	x	x	x	I	x	x	x	x	x	I					
P3	x	I	x	x	x	x	x	I	x	x	x	x	x	I	x	x	x	x	x	I				
P4	x	x	I	x	x	x	x	x	I	x	x	x	x	x	I	x	x	x	x	x	I			
P5	x	x	x	I	x	x	x	x	x	I	x	x	x	x	x	I	x	x	x	x	x	I		
P6	x	x	x	x	I	x	x	x	x	x	I	x	x	x	x	x	I	x	x	x	x	x	I	

SOURCE: Data from Bureau of Justice Statistics (2008b).

not used in estimation, but data from wave 1 have been used since that time with an adjustment to account for potential telescoping (see below, under Estimation).

DATA COLLECTION AND SURVEY MODE

As just noted, each address selected for the NCVS remains in the sample for 3 years, with seven interviews of household members taking place at 6-month intervals. The sample is address based, so that if a household or household member moves during the years of the panel, the interviews continue with the current household members who reside at the sampled address.

The initial contact and interview (wave 1) on the NCVS are conducted in person by a Census Bureau field representative. For subsequent interviews, the field representative may, and most often does, conduct the interview by telephone; personal interviews are conducted subsequently on the basis of a respondent's request and as needed to maintain continued response. Both in-person and telephone interviewing use computer-assisted survey instruments. Overall, 55 percent of person-level interviews in 2011 were conducted over the telephone.

The NCVS data collection protocol calls for the direct interviewing of each person 12 years or older in the household, with interviews held separately for each household member. Proxy interviewing instead of direct interviewing is permissible for minors (when a knowledgeable household member insists they not be interviewed directly); incapacitated persons; and individuals absent from the household during the entire period of field interviewing.

SURVEY INSTRUMENTS

Three survey instruments support the NCVS: the control card, the basic screen questionnaire, and the incident report.[11] Respondents are first screened to determine whether they experienced an incident in which they were victimized during the past 6-month period. In the second part of the interview, detailed information about any incident reported during the screening process is collected. The process is described in more detail below.

The control card is used to build a roster of household members, to determine eligibility to be interviewed, and to obtain basic information about the sample unit, including a record of visits, telephone calls, interviews, and reasons for any noninterview. It lists the name, age, gender, marital status,

[11]The survey instruments (both screening questionnaire and incident report) are displayed on the BJS website. Appendix C provides the URLs where these instruments can be found.

education, and relationships of all persons residing in a household. The importance of the control card lies in its brief description of all victimizations reported by household members in each interview. It serves as a quick reference during future interviews to ensure that victimizations previously counted are not reported again in error.

The basic screen questionnaire is used to screen for incidents that occurred during the 6-month reference period in which the individual household member may have been victimized. This is a very critical part of the interview (see "Cue Screening Questions" below).

An incident report is completed for each event recorded in the basic screen questionnaire. The report collects detailed information about each victimization, including the time, date, and place; information about the offender, including any relationship to the victim; circumstances of the incident, such as whether a weapon was used; consequences to the victim, including injury and loss of or damage to property; and whether the incident was reported to police. The incident report, along with the screening report, is subsequently used to categorize the incident as to whether it is a crime, and if so, what type of crime. This process is discussed in greater detail in Chapter 7.

Cue Screening Questions

During the screening phase, respondents are asked both directly and indirectly about potential victimizations through short cue screening questions. In an omnibus victimization survey like the NCVS, it would be difficult (and burdensome) to list each specific possible type of victimization in the screening questionnaire. Instead the screener contains cues designed to trigger memories of victimizations through cues related to particular contexts, locations, weapons, and offenders. A respondent may report any victimization at any point during the screening process. Several cue screening questions deal specifically with physical attacks, threats, or sexual activity.

> *Since [end date for 6-month reference period], were you attacked or threatened OR did you have something stolen from you: (a) at home including the porch or yard; (b) at or near a friend's, relative's, or neighbor's home; (c) at work or school; (d) in places such as a storage shed or laundry room, a shopping mall, restaurant, bank, or airport; (e) while riding in any vehicle; (f) on the street or in a parking lot; (g) at such places as a party, theater, gym, picnic area, bowling lanes, or while fishing or hunting; OR (h) did anyone attempt to attack or attempt to steal anything belonging to you from any of these places?*

> *(Other than any incidents already mentioned,) has anyone attacked or threatened you in any of these ways: (a) with any weapon, for instance, a gun or knife; (b) with anything like a baseball bat, frying pan, scissors, or stick; (c) by something thrown, such as a rock or bottle; (d) include any grabbing, punching, or choking; (e) any rape, attempted rape or other type of sexual attack; (f) any face-to-face threats; OR (g) any attack or threat or use of force by anyone at all? Please mention it even if you are not certain it was a crime.*
>
> *People often don't think of incidents committed by someone they know. (Other than any incidents already mentioned,) did you have something stolen from you OR were you attacked or threatened by (a) someone at work or school, (b) a neighbor or friend, (c) a relative or family member, (d) any other person you've met or known?*
>
> *Incidents involving forced or unwanted sexual acts are often difficult to talk about. (Other than any incidents already mentioned,) have you been forced or coerced to engage in unwanted sexual activity by (a) someone you didn't know before, (b) a casual acquaintance? OR (c) someone you know well?*
>
> *During the last 6 months [other than any incidents already mentioned), did you call the police to report something that happened to YOU which you thought was a crime?*

If the respondent replies *yes* to any of the cue screening questions, then he or she is asked to briefly describe in his or her own words what happened.

Incident Report

In the incident report, the respondent is further queried in detail about each incident reported in the screener. There are questions that relate to the location of the incident, weapon used (if any), injuries, medical care and expenses, distress as a victim, others present during the incident, etc. Listed below are questions that deal specifically with physical attacks, threats, or sexual activity.

> *Did the offender hit you, knock you down or actually attack you in any way? Did the offender TRY to attack you? Did the offender THREATEN you with harm in any way? What actually happened?*

If the respondent says an attack was attempted or threatened, then she or he is asked:

> *How did the offender TRY to attack you? How were you threatened? Any other way?*

If the respondent reports an unwanted sexual contact, then she or he is asked:

> *You mentioned some type of unwanted sexual contact with force. Do you mean forced or coerced sexual intercourse including attempts?*

If the respondent mentions rape, then he or she is asked:

> *You mentioned rape. Do you mean forced or coerced sexual intercourse?*

If the response is *no*, then he or she is asked:

> *What do you mean?*

The same queries are used for attempted rapes.

Post Processing

During processing, BJS staff, working with Census Bureau staff, classifies reported incidents into categories using a type-of-crime algorithm. Incidents involving sexual violence or unwanted activity are coded into one of several distinct categories: completed rape, attempted rape, sexual assault with serious assault, sexual assault with minor assault, sexual assault without injury, unwanted sexual contact without force, verbal threat of rape, and verbal threat of sexual assault. Figure 7-4 in Chapter 7 shows a flow chart of this process.

Even if the respondent reported "no" to the first query on rape in the cue questions, the respondent would still be classified as a rape victim if she or he reports a rape at any point in the incident report or at any time during the interview. Thus, classifying a respondent as a victim of rape or sexual assault is a two-stage process in the NCVS methodology, and it can occur through numerous avenues.

BJS uses the following definitions for rape and sexual assault (Bureau of Justice Statistics, n.d.-b):

Rape: Forced sexual intercourse including both psychological coercion as well as physical force. Forced sexual intercourse means vaginal, anal or oral penetration by the offender(s). This category also includes incidents where the penetration is from a foreign object such as a bottle. Includes attempted rapes, male as well as female victims, and both heterosexual and homosexual rape. Attempted rape includes verbal threats of rape.

Sexual assault: A wide range of victimizations, separate from rape or attempted rape. These crimes include attacks or attempted attacks generally involving unwanted sexual contact between victim and offender. Sexual assaults may or may not involve force and include such things as grabbing or fondling. Sexual assault also includes verbal threats.

ESTIMATION AND PRODUCTS

This section first briefly describes post collection weighting and adjustments in the NCVS. It then describes a recent change in the way that series victimization (repeated victimization of one victim) is handled in summarization. Lastly, the section briefly highlights some of the important public data products that come from the NCVS.

Estimation

The estimation procedures[12] for the NCVS begin with assigning a base weight to each selected unit, which is the reciprocal of the probability of the unit's selection for the sample. This base weight provides a rough measure of the population numbers represented by each unit in the sample. Next, an adjustment is made to the weight to account for nonresponse, which includes both a household nonresponse adjustment and a within-household adjustment. These adjustments are standard weighing procedures for household surveys.

BJS makes two additional adjustments. First, data from responders in a first interview (both for the household and each individual respondent) are adjusted to minimize the effect of telescoping on what is an unbounded interview.[13] The adjustment factor is calculated based on the following ratio. TIS stands for "time in sample." Thus TIS 1 refers to respondents who are

---

[12]The estimation procedures for the NCVS are described in *Survey Methodology for Criminal Victimization in the United States* (Bureau of Justice Statistics, 2008b). Unfortunately this documentation is out of date (2008), so the panel depended on several internal memoranda to complete its description of the estimation process.

[13]"Telescoping" is the tendency of respondents to report events that occurred prior to the start of the reference period. In this case, it would be victimizations that occurred more than 6 months ago. The initial interview then forms a "bound" that helps to minimize telescoping in subsequent interviews.

being interviewed for the first time, and TIS 2 refers to respondents who are being interviewed for the second time.

$$\text{Adjustment factor TIS 1} = \frac{\dfrac{\text{Weighted number of crimes}^{14}\text{ in TIS other than 1}}{\text{Weighted total of cases}^{15}\text{ in TIS other than 1}}}{\dfrac{\text{Weighted number of crimes in TIS 1}}{\text{Weighted total of cases in TIS 1}}}$$

That is, the potential overreporting of crimes in TIS 1 is adjusted to the average over the other times in sample.

A similar adjustment is made for initial interviews in other TIS groups, 2 through 6, denoted as *a* for reinstated sample cases:

$$\text{Adjustment factor TIS } a = \frac{\dfrac{\text{Weighted number of crimes in TIS } a}{\text{Weighted total of cases in TIS } a}}{\dfrac{\text{Weighted number of crimes in TIS 1}}{\text{Weighted total of cases in TIS 1}}}$$

The final weight also includes ratio adjustments to known population totals based on the adjusted counts from the most recent census (currently 2010). The adjustments are applied at two stages:

- In stage 1, ratio adjustments are applied to the data on the basis of PSU-level estimates of individuals by race and zone of residence to more closely match the census PSU estimates for those population totals. PSUs that are included with certainty (self-representing) receive an adjustment ratio of 1.
- In stage 2, ratio adjustments are applied to the weighted counts of individuals from the NCVS to census estimates (adjusted for the undercount) of population totals by sex, race, and age categories.

Thus, for estimation, the final person weight is the product of the values of the base weight and the adjustments described above. The final household weight is the product of all components except the within-household noninterview adjustment component.

[14]Criminal victimizations.

[15]Respondents.

Annual estimates of the levels and rates of victimization are derived by accumulating four quarterly estimates (which are not publicly available). The weights of all crimes reported during interviews in that year are summed, regardless of when the crime occurred. The base weight for personal crime is the sum of all person weights.

Series Victimization

Series victimization is a category used when a respondent reports that six or more separate but similar victimizations have occurred during the 6-month reference period, but is unable to recall enough details of each incident to distinguish them from one another. Until recently, the NCVS estimates published in *Criminal Victimization* excluded series victimizations (or included them as a single victimization in certain special reports), which clearly undercounted the total number of all types of victimizations, including rape and sexual assault.

Beginning with *Criminal Victimizations, 2011*, BJS began including series victimizations directly in its estimates. The NCVS uses the victim's report of the number of similar victimizations, with a maximum of 10, and collects (and applies to each victimization) detailed information only for the most recent victimization. These new procedures are being applied to all types of victimizations, including rape and sexual assault (Bureau of Justice Statistics, 2012a, p. 13):

> BJS now includes series victimizations using the victim's estimate of the number of times the victimizations occurred over the past 6 months, capping the number of victimizations within each series at a maximum of 10. This strategy for counting series victimizations balances the desire to estimate national rates and account for the experience of persons with repeat victimizations while noting that some estimation errors exist in the number of times these victimizations occurred. This bulletin is the first to include series victimizations throughout the entire report, and all victimizations estimates in this report reflect this new count strategy.

A technical report provides findings on the extent and nature of series victimization (Lauritsen et al., 2012, p. iii):

> Including series victimizations in national rates results in rather large increases in the level of violent victimizations; however, trends in violence are generally similar regardless of whether series victimizations are included. The impact of including series victimizations may vary across years and crime types, in part reflecting the relative rarity of the offense type under consideration.

BJS has revised estimates back to 1993 in its online database, accessible through the NCVS victimization analysis tool.[16] The effects of this change on the estimates of rape and sexual assault are so substantial that the panel decided to include both estimates—one based on the new method of including up to 10 victimizations in series, and one in which series victimizations are excluded—in our comparisons in Chapters 6 and 7. The revised estimation process means that a very small number of reports have a major impact on the estimates. (This issue is further discussed in Chapters 6, 7, and 10.)

Products and Statistics

The NCVS provides a major database of information about criminal victimizations. BJS's major annual publication is *Criminal Victimization*, which includes estimates of rape and sexual assault. Figure 4-1 and Table 4-4 provide the historical trend for the number of victimizations, which shows a steady downward direction with considerable year-to-year instability.[17] As explained above, the estimates that include series victimizations and the estimates excluding these (in Table 4-4) are substantially different.

The NCVS has three primary rotating supplements that are conducted on a periodic basis: on school crime, identity theft, and on police-public contact. BJS also pools NCVS data across years to produce special reports, such as *Female Victims of Sexual Violence, 1994-2010* (Planty et al., 2013).

In addition to its publications, BJS assists and encourages researchers to undertake their own analysis using the online NCVS victimization analysis tool (see footnote 16), which allows data users to produce custom tables. BJS also offers a visiting fellows program, which broadens the collaboration between the agency and academic scholars. Data users are able to conduct research in conjunction with BJS researchers and staff and have access to facilities and microdata[18] for that research. BJS also supports the archiving of NCVS data in the National Archive of Criminal Justice Data that is part of the Inter-university Consortium for Political and Social Research (ICPSR).

---

[16]The tool is available at http://bjs.gov/index.cfm?ty=nvat [May 2013].

[17]It is useful to note that the decline in rape and sexual assault victimizations between 1993 and 2011 is consistent with the pattern of change in all violent crime during these years. The number of all violent crime victimizations in 2011 was only 35 percent of the number estimated for 1993 (calculated using the NCVS victimization analysis tool, Bureau of Justice Statistics, n.d.-a). This trend appears in the United States and in other industrialized nations.

[18]In this context, microdata are data provided at the individual respondent level rather than summarized data.

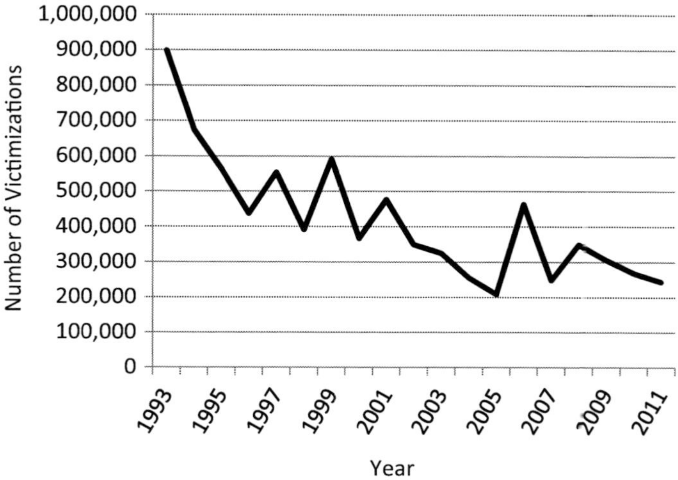

FIGURE 4-1 Number of victimizations of rape and sexual assault estimated by the National Crime Victimization Survey (NCVS) (including series victimization).
SOURCE: Data from the NCVS victimization analysis tool, Bureau of Justice Statistics (n.d.-a).

TABLE 4-4 Statistics on Rape and Sexual Assault Criminal Victimizations, NCVS, 1990-2011

	Series Criminal Victimization Excluded			
Year	Number of Criminal Victimizations	Standard Error for Number of Criminal Victimizations	Criminal Victimization Rate per 1,000 People, Aged 12+	Standard Error for Rate
1990	130,260	22,036[a]	0.6	NA
1991	173,000	28,009	0.8	.13
1992	Transition year	NA	NA	NA
1993	521,223[b]	51,573	2.5[b]	.24
1994	443,509[b]	36,858	2.1[b]	.17
1995	363,527[b]	32,830	1.7[b]	.15
1996	307,100	30,430	1.4	.14
1997	311,110	34,409	1.4	.15
1998	332,500	36,103	1.5	.16
1999	383,170	38,340	1.7	.17
2000	260,950	32,152	1.2	.14
2001	248,250	31,043	1.1	.13
2002	247,730	31,352	1.1	.14
2003	198,850	27,884	0.8	.12
2004	209,880	29,887	0.9	.12
2005	190,592[c]	31,032	0.8	.13
2006	260,940	36,990	1.1	.15
2007	248,277	32,924	1.0	.13
2008	203,830	31,719	0.8	.13
2009	125,910	24,079	0.5	.09
2010	188,380	29,399	0.7	.11
2011	217,331	32,616	0.8	.13

[a]Calculated based on formula provided in Bureau of Justice Statistics (BJS) *Criminal Victimization, 1990.*

[b]The estimates published in BJS *Criminal Victimization, 1993, 1994, 1995* were revised in 1996 to reflect a methodology change to estimate victimizations for the "collection year" rather than the year in which the victimization occurred.

[c]Based on errata issued by BJS on June 16, 2011.

SOURCES: Bureau of Justice Statistics (1991, 1992, 1993, 1994, 1995, 1996, 1997, 1998, 1999, 2000, 2001, 2002a, 2003, 2004, 2005, 2006, 2007, 2008a, 2009, 2010, 2011, 2012a, n.d.-a).

Series Criminal Victimization Counted up to Ten Incidents

Number of Criminal Victimizations	Standard Error for Number of Criminal Victimizations	Criminal Victimization Rate per 1,000 People, Aged 12+	Standard Error for Rate
NA			
NA			
NA			
898,239	71,348	4.3	.33
674,291	47,198	3.2	.22
563,249	42,418	2.6	.20
437,198	37,345	2.0	.17
553,523	48,035	2.5	.21
391,101	39,935	1.8	.18
591,460	49,907	2.6	.22
366,747	39,485	1.6	.17
476,578	46,216	2.1	.20
349,805	38,253	1.5	.17
325,311	36,759	1.4	.15
255,769	33,339	1.1	.14
207,760	32,551	0.8	.13
463,598	50,305	1.9	.20
248,277	32,924	1.0	.13
349,691	42,837	1.4	.17
305,574	39,443	1.2	.16
268,574	36,057	1.0	.14
243,803	34,800	0.9	.14

5

Selected Other Surveys on Rape and Sexual Assault

Chapter 1 describes two perspectives for measuring rape and sexual assault through surveys, the criminal justice perspective, and the public health perspective. The first perspective focuses on measuring criminal victimizations as exemplified in the National Crime Victimization Survey (NCVS), described in Chapter 4. It provides counts of criminal victimizations in a specified period of time, calculates the victimization rates for that period, and calculates the change in rate from a previous period. The second perspective, public health, focuses more broadly on sexual violence and the effects of that violence on the physical and emotional health of its victims.

The panel believes that the two approaches have some basic differences in purpose, and this purpose differential has led to certain methodological decisions as the surveys were designed. The report maintains this dichotomy to help clearly explain these conceptual frameworks. As the report proceeds, the panel concludes that some of the design decisions in the NCVS might be improved by adopting some of the methodological approaches used by the public health approach.

This report does not try to make recommendations about how the surveys designed under the public health approach might be improved. With limited time and resources, the panel made the decision to focus its analysis on the NCVS with the intent to make specific recommendations to the Bureau of Justice Statistics (BJS) for future estimation of rape and sexual assault. Two of the public health surveys are more than 16 years old and one is almost 25 years old. Thus it was a low priority to try to recommend ways these older surveys could have been improved, and it would have been very difficult to obtain the needed metadata to make those recommendations. The panel believes that a thorough error analysis of the National Intimate Partner and Sexual Violence Survey (NISVS) would be useful because it is likely to be conducted on a continuing basis. The panel hopes that the Centers for Disease Control and Prevention (CDC) will commission such an analysis.

The two approaches differ in terms of goals and survey methodologies. For the NCVS, there are two key elements: the classification of a victimization as criminal or noncriminal (according to criteria from the BJS), and the determination of whether the victimization occurred within a specified reference period. In the public health approach, those elements are less important; instead, the key focus is trying to measure the effect of victimizations. However, surveys based on the public health perspective still provide measures (often "lifetime measures") of rape and sexual assault, as discussed below. Because of the basic conceptual difference in their approach, their estimated victimization rates would be expected to be and are different from those in the NCVS.

This chapter provides an overview of four significant surveys that generally fall under the public health approach: the National Women's Study (NWS), the National Violence Against Women Study (NVAWS), the National College Women Sexual Victimization (NCWSV) Study, and the National Intimate Partner and Sexual Violence Survey (NISVS). This chapter, like Chapter 4, does not try to assess the quality of these surveys; rather, it provides summary descriptions of the four surveys. By including these specific surveys in this descriptive chapter, the panel is not suggesting that the estimates from these surveys represent truth or are somehow a "gold standard." Nor did we include all relevant studies of rape and sexual assault.[1] Nevertheless, this selection provides an overview of other work done in this area. Chapter 6 subsequently provides some reasonable comparison among the surveys, and with the NCVS and the Uniform Crime Reports (UCR) Program.

NATIONAL WOMEN'S STUDY (1989-1991)

The NWS, funded by the National Institute of Drug Abuse, was one of the first of surveys that provided national-level measures of rape and sexual assault from the public health perspective. It was conducted in 1989-1991 with a national sample of 4,008 women. The results were published in

[1]The reader may be aware of additional studies that are relevant to these discussions. The panel identified four studies that it describes in this chapter, and Appendix D provides more details on these four studies and includes three other studies not discussed in the report proper. The panel did not attempt to make a comprehensive list of such studies because it needed to keep its focus on the examination of the NCVS.

Rape in America: Report to the Nation (Kilpatrick, Edmunds, and Seymour, 1992).[2]

Methodology

The NWS was a longitudinal survey of women aged 18 years of age and older, divided into two parts: a sample of all adult women (sample size of 2,008) and an oversample of younger women, those between the ages of 18 and 34 (sample size of 2,000). The households were selected using a two-stage area probability sampling procedure. In the first stage, the United States was divided into four geographic regions and three census-based size-of-place strata, which resulted in 12 mutually exclusive and exhaustive groupings of the total U.S. geographic area. In the second stage, random digit dialing (RDD) was used to select households within each geographic area. After reaching a household, the interviewer ascertained the number of adult females in the household and randomly selected one for an interview.

Data collection was conducted in three waves, following respondents from wave 1 through two additional waves. The survey protocol used only female interviewers. All three waves assessed women's lifetime experience of forcible rape. Wave 2 assessed forcible rapes that occurred between the baseline and second interview, while wave 3 assessed forcible rapes that occurred between the second and third interviews. Wave 1 collected information about the lifetime prevalence of rape, along with descriptive information about rape incidents.[3] Wave 2 was conducted 1 year later and asked respondents about prevalence of rape during the previous year. Thus, the 12-month recall questions were bounded by the wave 1 interview. Wave 3 collected information about rape victims' concerns, medical examinations, willingness to report future rapes to police, and opinions about the impact of protection of their names from media disclosure. Potential mental health problems were assessed during each of the three waves.

Wave 1 was completed with a cooperation rate of 85 percent and a response rate of 34 percent.[4] For wave 2, 81 percent of the wave 1 partici-

[2]Dean Kilpatrick was the principal investigator on this study. Sample selection and survey operations were performed by the firm of Schulman, Ronca, and Bucuvalas, Inc. (SRBI), a survey research organization based in New York City.

[3]The NWS also included some screening questions for other forms of contact and noncontact pressure for sexual activity (see Resnick et al., 1993).

[4]The cooperation rate (percentage of those contacted that agreed to cooperate) was calculated as a part of the original report and is documented there. The panel used the original case dispositions to calculate a response rate using Standard 4 of the American Association of Public Opinion Research (n.d.). Cooperation rates are generally calculated once a potential respondent has been contacted and do not reflect sample screen outs that occur before a respondent is contacted.

pants were located and reinterviewed. The data collected were weighted to align with Census Bureau projections of the number of adult females by age and race.

The NWS questionnaire was different from the one used in the National Crime Survey (NCS)[5] in that it contained explicitly worded questions about sexual intercourse. It also included questions about oral and anal rape. The NWS also attempted to look at major post-incident mental health problems such as depression, post-traumatic stress disorder, suicide attempts, and alcohol- and drug-related problems.

Questions

Given its public health context and perspective, the survey included a broad range of questions (Kilpatrick, Edmunds, and Seymour, 1992, p. 1):

> In addition to gathering information about forcible rapes that occurred throughout women's lifetimes, the National Women's Study also assessed such major mental health problems as depression, Post-traumatic Stress Disorder, suicide attempts, as well as alcohol and drug-related problems and consumption.

Unlike the NCVS, the NWS did not use separate screening and incident reports. The following questions were used to capture incidents of rape in the NWS (Kilpatrick, Edmunds, and Seymour, 1992, p. 15):

> *Women do not always report such experiences to police or discuss them with family or friends. The person making the advances isn't always a stranger, but can be a friend, boyfriend, or even a family member. Such experiences can occur anytime in a woman's life—even as a child. Regardless of how long ago it happened or who made the advances. . .*
>
> - *Has a man or boy ever made you have sex by using force or threatening to harm you or someone close to you? Just so there is no mistake, by sex we mean putting a penis in your vagina.*
> - *Has anyone ever made you have oral sex by force or threat of harm? Just so there is no mistake, by oral sex, we mean that a man or boy put his penis in your mouth or somebody penetrated your vagina or anus with his mouth or tongue.*
> - *Has anyone ever made you have anal sex by force or threat of harm?*

[5]The NCS was the predecessor to the current NCVS; see Chapters 1 and 4.

- *Has anyone ever put fingers or objects in your vagina or anus against your will by using force or threat?*

Results

The NWS estimated prevalence of rape victimization rather than estimating the number of incidences of rape as was measured by the NCS. (Attempted rape and other types of sexual assault were measured in the survey but these estimates were not included in the report.) The NWS prevalence estimate was 683,000 adult females per year who were raped in the United States. This total was 5 times higher than the number of incidences estimated that same year by the NCS (130,000, which included rape, attempted rape, and other sexual assaults) and almost seven times the number of incidences summarized by the UCR system (102,560, which included attempted rape but not other types of sexual assault) (see Figure 5-1). The NWS was one of the first major surveys that provided evidence of undercounting of rape on both the UCR and the NCS.

NATIONAL VIOLENCE AGAINST WOMEN STUDY (1995-1996)

In the Violence Against Women Act of 1994,[6] Congress mandated that the federal government provide a more valid estimate of the magnitude of violence against women, including both rape and stalking. The National Institute of Justice (NIJ) and the CDC partnered, through a grant to the Center for Policy Research, to launch a national survey. The survey was fielded once in 1995 and became known as the NVAWS (see Tjaden and Thoennes, 2000).[7]

Methodology

Like the NWS, the NVAWS was designed around an RDD telephone survey. It was conducted once, from November 1995 to May 1996, targeting adults, both male and female, covering all households with a landline telephone in 50 states and the District of Columbia. The sample was administered within the U.S. Census Bureau regions. Interviewers called RDD-selected numbers from a central telephone facility using computer-assisted technology. Nonworking and nonresidential numbers were screened out.

[6]The act required the Attorney General to report on the incidence of violence against women including stalking (P.L. 103-322, Section 40610). The act was a part of the larger Omnibus 1994 Crime Control Act.

[7]Patricia Tjaden and Nancy Thoennes were the principal investigators on the study. Sample selection and survey operations were performed by SRBI, a survey research organization based in New York City.

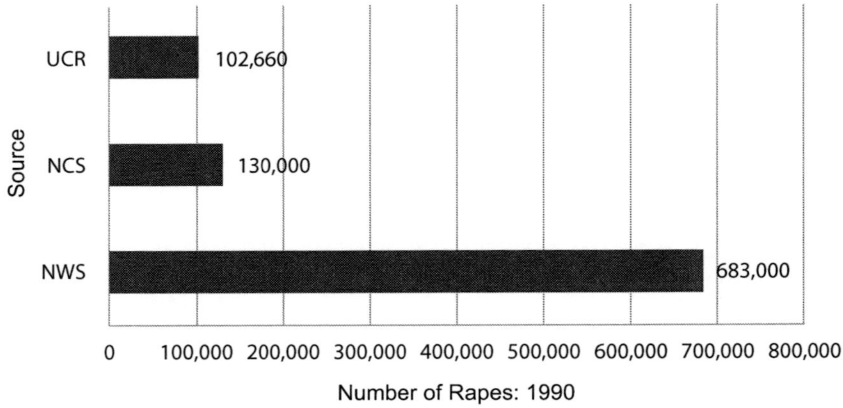

FIGURE 5-1 Rapes of females: results from three data sources.
NOTES: The data are from the National Women's Study (NWS), the National Crime Study (NCS; predecessor of the NCVS), and the Uniform Crime Reports (UCR). The NWS estimate is for prevalence of completed rapes (raped at least one time) for females who were 18 years of age and older. The NCS estimates the number of rapes and attempted rapes for females who were 12 years of age and older. The UCR estimate is for both completed and attempted rapes reported to the police, no specific age limit.
SOURCE: Kilpatrick, Edmunds, and Seymour (1992, p. 3, Figure 2).

Once a residential household was reached, all eligible adults (i.e., women and men 18 years of age and older) in each household were identified. In households with more than one eligible adult, the adult with the most recent birthday was selected as the designated respondent.

A total of 8,000 women and 8,005[8] men were interviewed. All female respondents were interviewed by female interviewers. For male respondents, approximately half of the interviews were conducted by female interviewers and half by male interviewers. Bilingual interviewers administered a Spanish-language version of the questionnaire for Spanish-speaking respondents. Because the survey was conducted only once, there was no initial interview that bounded the 12-month reporting of victimizations. The survey had a 72 percent cooperation rate for females and a 69 percent cooperation rate for male respondents.[9] The unweighted response rate,

[8]Five completed interviews were subsequently eliminated from the data file during editing because of an excessive amount of incongruous data.

[9]The cooperation rate (percentage of contacted individuals who responded) for the NVAWS was calculated by dividing the number of completed interviews (including those that were screened out because they were ineligible) by the total number of completed interviews, screened-out interviews, refusals, and terminated interviews (Tjaden and Thoennes, 2000, p. 4).

calculated based on American Association for Public Opinion Research (AAPOR) Standard 4, was 34 percent (see footnote 4).

Questions

The NVAWS followed the NWS approach to question formulation, using behaviorally specific words to ask questions about rape. The survey, too, also covered a fairly wide range of topics that included not only rape and attempted rape, but also physical assault experienced as a child by adult caretakers, physical assault experienced as an adult, and stalking. It asked for "detailed information about the characteristics and consequences of victimization for each type of perpetrator identified by the respondent" (Tjaden and Thoennes 2000, p. 1).

The following specific questions were asked about rape and attempted rape (Tjaden and Thoennes, 2000, p. 4):

1. *Has a man or boy ever made you have sex by using force or threatening to harm you or someone close to you? Just so there is no mistake, by sex we mean putting a penis in your vagina.*
2. *Has anyone, male or female, ever made you have oral sex by using force or threat of force? Just so there is no mistake, by oral sex we mean that a man or boy put his penis in your mouth or someone, male or female, penetrated your vagina or anus with their mouth.*
3. *Has anyone ever made you have anal sex by using force or threat of harm? Just so there is no mistake, by anal sex we mean that a man or boy put his penis in your anus.*
4. *Has anyone, male or female, ever put fingers or objects in your vagina or anus against your will or by using force or threats?*
5. *Has anyone, male or female, ever attempted to make you have vaginal, oral, or anal sex against your will, but intercourse or penetration did not occur?*

The NVAWS used a single-stage classification process to identify rape victims, unlike the two-stage process currently used in the NCVS (see Chapter 4). If a respondent responded *yes* to one of the above questions, then he or she was classified as a victim of a completed or attempted rape (depending on the question).

Results

The survey generated lifetime and 12-month prevalence rates, as well as 12-month incidence rates.[10] The survey estimated that 302,091 women and 92,748 men had been raped in the previous 12 months (prevalence), with a total of 987,362 rape victimizations. The victims were women in 89 percent (876,064) of the incidents. The estimated total 12-month incidence rate was 5.1 per 1,000 people (18+ years). But the NVAWS sample of 16,000 women and men interviewed included only 24 women and 8 men who reported having been raped in the past 12 months—a low incidence. Thus, the relative standard errors for incidence rates and for male 12-month prevalence rates are large, above 30 percent (see Table 5-1 for details).

NATIONAL COLLEGE WOMEN SEXUAL VICTIMIZATION STUDY (1997)

Sexual violence has been, and continues to be, a particular problem on college campuses (Diamond and Emerson, 2012; Fisher, Cullen, and Turner, 2000; Karjane, Fisher, and Cullen, 2005). In the Student Right-to-Know and Campus Security Act of 1990 (P.L. 101-542), Congress expressed concern about the seriousness of sexual violence on college campuses and highlighted the importance of having accurate information about such violence. The act mandated that colleges and universities participating in federal student aid programs "prepare, publish, and distribute, through appropriate publications or mailings, to all current students and employees, and to any applicant for enrollment or employment upon request, an annual security report" (20 USC Section 1092). The report was required to include campus security policies and campus crime statistics.

The NCWSV (Fisher, Cullen, and Turner, 2000) was a response to those general concerns and the specific legislation. It was funded by the National Institute of Justice of the U.S. Department of Justice and conducted once, in 1997.[11]

Methodology

The NCWSV was based on a national sample of 4,446 women who were attending 2- or 4-year colleges or universities during fall 1996. The design had two stages: sampling institutions and then sampling female

---

[10]An incidence rate is based on the total number of individual victimizations that occurred during the reference period. A prevalence rate is based on the total number of victims that experienced one or more victimizations during the reference period.

[11]The principal investigators were Bonnie Fisher and Francis Cullen. SRBI, a survey research organization based in New York City, conducted the survey.

TABLE 5-1 Estimated Number of Rape, Physical Assault, and Stalking Victimizations Annually by Gender of Victim, National Violence Against Women Study

Type of Victimization	Percentage of Sample Reporting an Incident	Estimated Number of Victims	Average Number of Victimizations per Victim[a]	Estimated Number of Victimizations	Annual Rate of Victimization per 1,000 Persons
Women, 18+ (100,697,000) sample size = 8,000					
Rape	0.3	302,091	2.9[b]	876,064[b]	8.7
Physical assault	1.9	1,913,243	3.1	5,931,053	58.9
Stalking	1.0	1,006,970	1.0	1,006,970	10.0
Men, 18+ (92,748,000) sample size = 8,000					
Rape	0.1	92,748	1.2[b]	111,298[b]	1.2
Physical assault	3.4	3,153,432	2.5	7,883,580	85.0
Stalking	0.4	370,992	1.0	370,992	4.0
Total,[c] 18+ (193,445,000) sample size = 16,000					
Rape	0.2	394,839	2.5	987,362	5.1

[a]The standard error of the mean is 1.4 for female rape victims, 0.2 for female physical assault victims, 0.5 for male rape victims, and 0.2 for male physical assault victims.

[b]Relative standard error exceeds 30 percent.

[c]Calculated from statistics in table about males and females.

SOURCE: National Violence Against Women Study (Tjaden and Thoennes, 2006, p. 15).

students in the selected institutions. The population of eligible institutions consisted of schools with at least 1,000 students. The list of institutions was stratified by the size of the total student enrollment (1,000-2,499; 2,500-4,999; 5,000-19,999; 20,000 or more) and the school's location (urban, suburban, and rural). Schools were randomly chosen using a probability proportional to the total female enrollment. For each selected institution, the American Student List Company provided a list of female students attending the institution in the fall of 1996. (The company also provided the school address and telephone number for each selected student.) Using this as a sampling frame, a random sample of female students was selected.

The survey was conducted between February and May 1997. Each sampled student was sent a presurvey letter 2 weeks prior to a telephone contact by a female interviewer, who called from a centralized facility. The interviewers used a computer-assisted telephone interviewing system (CATI). The response rate was 67.1 percent;[12] the reported cooperation rate was 85.6 percent.

To limit potential telescoping, a reference period for recall was established that would have a clear starting point for those students—beginning of school in fall 1996. Because the interviews were conducted the following February through May, the reference period was approximately 7 months long.

The NCWSV measured sexual victimization using the two-stage measurement format of the NCVS (see Chapter 4), which involves a screener questionnaire followed by a detailed incident report, but the screener's questions were very different from those in the NCVS. Drawing from both the NWS and the NVAWS, the NCWSV's screener questionnaire contained 10 behaviorally specific screen questions that sought to assess whether respondents had experienced a range of sexual victimizations (Fisher, Cullen, and Turner, 2000, p. 6):

1. *Since school began in the fall 1996, has anyone made you have sexual intercourse by using force or threatening to harm you or someone close to you? Just so there is no mistake, by intercourse I mean putting a penis in your vagina.*
2. *Since school began in the fall 1996, has anyone made you have oral sex by force or threat of harm? By oral sex, I mean did someone's mouth or tongue make contact with your vagina or anus or did your mouth or tongue make contact with someone else's genitals or anus.*
3. *Since school began in the fall 1996, has anyone made you have anal sex by force or threat of harm? By anal sex, I mean putting a penis in your anus or rectum?*

[12]The panel recalculated the response rate on the basis of AAPOR Standard 4: see footnote 4.

4. *Since school began in the fall 1996, has anyone ever used force or threat of harm to sexually penetrate you with a foreign object? By this, I mean, for example, placing a bottle or finger in your vagina or anus?*
5. *Since school began in fall 1996, has anyone attempted but not succeeded in making you take part in any of the unwanted sexual experiences that I have just asked you about? For example, did anyone threaten or try but not succeed to have vaginal, oral, or anal sex with you or try unsuccessfully to penetrate your vagina or anus with a foreign object or finger?*
6. *Not counting the types of sexual contact already mentioned, have you experienced any unwanted or uninvited touching of a sexual nature since school began in fall 1996? This includes forced kissing, touching of private parts, grabbing, fondling, and rubbing up against you in a sexual way, even if it is over your clothes.*
7. *Since school began in fall 1996, has anyone attempted but not succeeded in unwanted or uninvited touching of a sexual nature?*
8. *Since school began in fall 1996, has anyone made or tried to make you have sexual intercourse or sexual contact when you did not want to by making threats of nonphysical punishment, such as lowering a grade, being demoted or fired from a job, damaging your reputation, or being excluded from a group for failure to comply with requests for any type of sexual activity?*
9. *Since school began in fall 1996, has anyone made or tried to make you have sexual intercourse or sexual contact when you did not want to by promises of rewards, such as raising a grade, being hired or promoted, being given a ride or class notes, or getting help with coursework from a fellow student if you complied sexually?*
10. *Since school began in fall 1996, has anyone made or tried to make you have sexual intercourse or sexual contact when you did not want to by simply being overwhelmed by someone's continual pestering and verbal pressure?*

The NCWSV used these screening questions and subsequent incident reports to categorize and measure 12 types of sexual victimizations (see Table 5-2). In addition to the victimization measures, other questions covered a range of factors including stalking, respondents' demographic characteristics, lifestyle, routine activities, living arrangements, and prior sexual victimizations. Secondary data sources were used to measure the characteristics of the schools the respondents attended (e.g., size of enrollment, location, crime rate).

TABLE 5-2 Types of Victimizations Measured on the National College Women Sexual Victimization Study

Type of Victimization	Definition
Completed rape	Unwanted completed penetration by force or the threat of force. Penetration includes penile-vaginal, mouth on your genitals, mouth on someone else's genitals, penile-anal, digital-vaginal, digital-anal, object-vaginal, and object-anal.
Attempted rape	Unwanted attempted penetration by force or the threat of force. Penetration includes penile-vaginal, mouth on your genitals, mouth on someone else's genitals, penile-anal, digital-vaginal, digital-anal, object-vaginal, and object-anal.
Completed sexual coercion	Unwanted completed penetration with the threat of non-physical punishment, promise of reward, or pestering/verbal pressure. Penetration includes penile-vaginal, mouth on your genitals, mouth on someone else's genitals, penile-anal, digital-vaginal, digital-anal, object-vaginal, and object-anal.
Attempted sexual coercion	Unwanted attempted penetration with the threat of non-physical punishment, promise of reward, or pestering/verbal pressure. Penetration includes penile-vaginal, mouth on your genitals, mouth on someone else's genitals, penile-anal, digital-vaginal, digital-anal, object-vaginal, and object-anal.
Completed sexual contact with force or threat of force	Unwanted completed sexual contact (not penetration) with force or threat of force. Sexual contact includes touching; grabbing or fondling of breasts, buttocks, or genitals, either under or over your clothes; kissing; licking or sucking; or some other form of unwanted sexual contact.
Completed sexual contact without force	Any type of unwanted completed sexual contact (not penetration) with the threat of nonphysical punishment, promise of reward, or pestering/verbal pressure. Sexual contact includes touching; grabbing or fondling of breasts, buttocks, or genitals, either under or over your clothes; kissing; licking or sucking; or some other form of unwanted sexual contact.
Attempted sexual contact with force or threat of force	Unwanted attempted sexual contact (not penetration) with force or threat of force. Sexual contact includes touching; grabbing or fondling of breasts, buttocks, or genitals, either under or over your clothes; kissing; licking or sucking; or some other form of unwanted sexual contact.

SOURCE: Fisher, Cullen, and Turner (2000, Exhibit 2).

Results

Respondents in the study reported 123 incidents of rape or attempted rape, leading to the estimates of rape prevalence and incidents of female college students shown in Table 5-3. The prevalence rate for rape and attempted rape was 27.7 per 1,000 female students, and the incidence rate was 35.3 per 1,000 female students. It is important to point out that both of the rates are based on a reference period of approximately 7 months (academic year) and are not an annual rate. The authors state that "projecting results beyond this reference period is problematic for a number of reasons, such as assuming that the risk of victimization is the same during summer months and remains stable over a person's time in college" (Fisher, Cullen, and Turner, 2000, p. 10). However, it is likely that these rates would be higher than those published if they had accounted for an entire year rather than just 7 months.

To better understand how these results compared with those collected through the NCVS, Fisher and Cullen (1999) worked with BJS to run a comparison study in the 1996-1997 academic year that was conducted close enough in time to the NCWSV and used a similar sample design so that the results could be compared to the NCWSV. The sample size for this study was 4,432 college women. This study was designed to mimic the NCWSV with the same sampling methodology, contact protocol, and interviewers except that the wording on the screen questions and incident reports were aligned to those used on the NCVS (tailored to a college population). Thus the NCWSV used behaviorally specific wording and the

TABLE 5-3 Rape and Attempted Rape of Female College Students, NCWSV, 1996

	Victims			Incidents	
Type of Victimization	Number of Victims in the Sample	Percentage of Sample	Rate per 1,000 Female Students	Number of Incidents	Rate per 1,000 Female Students
Completed rape	74	1.7	16.6	86	19.3
Attempted rape	49	1.1	11.0	71	16.0
Total	123	2.8	27.7*	157	35.3

*Total has been rounded.
SOURCE: Fisher, Cullen, and Turner (2000, Exhibit 3, p. 11).

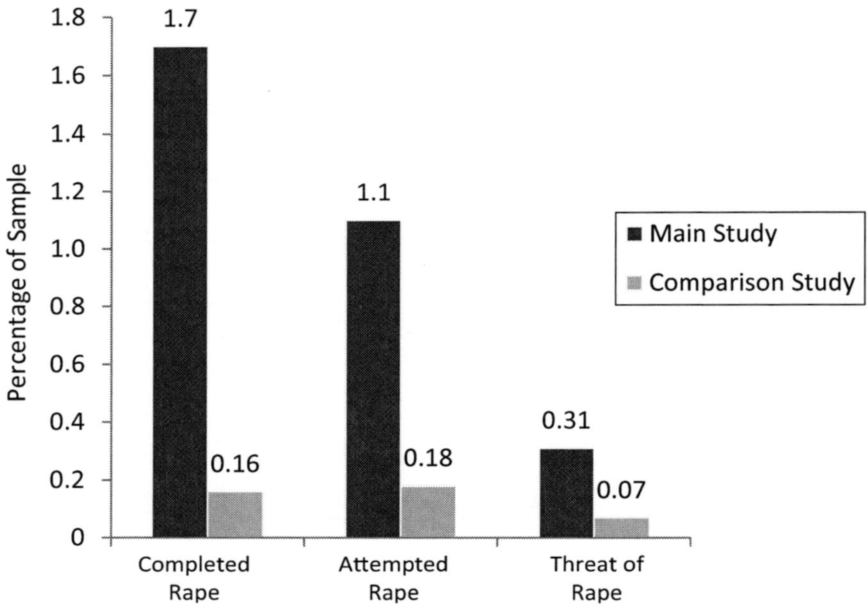

FIGURE 5-2 Comparisons of rape estimates between the National College Women Sexual Victimization Study main study and a comparison study. See text for discussion.
SOURCE: Data from Fisher (2009, p. 142).

BJS-sponsored comparison study did not.[13] Figure 5-2 provides results from this comparison. In each category—completed rape, attempted rape, and threat of rape—the main study (featuring behaviorally specific wording) resulted in more reports of incidents than did the comparison study (using NCVS wording). There was the greatest difference between the two studies for rape, somewhat less of a difference for attempted rape, and considerably less difference for the threat of rape. See also Fisher (2009).

NATIONAL INTIMATE PARTNER AND SEXUAL VIOLENCE SURVEY (2010)

Fifteen years after the NVAWS, the CDC and NIJ again partnered to fund the NISVS, with additional support from the Department of Defense.

[13]Fisher, Cullen, and Turner (2000, p.12) explain that answers in the comparison study "were not adjusted using verbatim responses [as is done with the NCVS]. We do not know how much this consideration affects the findings reported for the comparison component that is again, based on NCVS methodology."

It is a nationally representative survey that assesses experiences of sexual violence, stalking, and intimate partner physical violence, expressive aggression, and control among English and Spanish-speaking women and men, aged 18 years of age and older. This study was first fielded in 2010, and the CDC intends to conduct it on an annual basis.

The NISVS measures both 12-month and lifetime prevalence rates for the specified types of violence. This survey has a public health focus with concern that "unlike most other crimes, intimate partner violence or domestic violence is usually not a sudden, isolated, and unexpected incident. It may involve years of emotional and psychological trauma as well as physical injuries, which may become increasingly more severe and occur frequently over time" (Office for Victims of Crime, n.d.).

Methodology

The data collection for the NISVS was conducted by RTI International for the CDC. Like the sample design for the NVAWS and the NCWSV, discussed above, the NISVS sample design uses RDD technology to reach the target population. Unlike the other two surveys, however, the sampling frame for this study includes both landline and cell phones.

The first survey was conducted in 50 states and the District of Columbia from January 22 through December 31, 2010. A total of 18,049 interviews were conducted (9,970 women and 8,079 men) targeting the U.S. noninstitutionalized population aged 18 years of age and older. This includes 16,507 completed and 1,542 partially completed interviews. A total of 9,086 females and 7,421 males completed the survey. Approximately 45.2 percent of interviews were conducted from the landline telephone frame and 54.8 percent of interviews were conducted from the cell phone frame. Advance letters were sent to approximately 50 percent of the landline sample addresses (obtained by using reverse address matching to the telephone numbers). The survey used only female interviewers. The overall weighted response rate in NISVS ranged from 27.5 to 33.6 percent.[14] The cooperation rate was 81.3 percent.

The NISVS 2010 Summary Report included estimates for five different categories of sexual victimizations including completed and attempted rape (see Box 5-1). Questions used to measure 12-month prevalence were unbounded by a previous survey or event. Respondents were first asked

[14]This range in response rates reflects the differences in how the proportion of unknowns that might have been eligible but not interviewed is estimated, as applied in AAPOR response rate computation standards. These variations each handle "unknowns" (phone numbers that were never answered) differently. Assumptions are made based on respondents with known eligibility status for the survey.

BOX 5-1
Sexual Violence Defined by the NISVS

Five types of sexual violence were measured in NISVS. These include acts of rape (forced penetration), and types of sexual violence other than rape.

Rape is defined as any **completed or attempted** unwanted vaginal (for women), oral, or anal penetration through the use of physical force (such as being pinned or held down, or by the use of violence) or threats to physically harm and includes times when the victim was drunk, high, drugged, or passed out and unable to consent. Rape is separated into three types, completed forced penetration, attempted forced penetration, and completed alcohol or drug facilitated penetration.

- Among women, rape includes vaginal, oral, or anal penetration by a male using his penis. It also includes vaginal or anal penetration by a male or female using their fingers or an object.
- Among men, rape includes oral or anal penetration by a male using his penis. It also includes anal penetration by a male or female using their fingers or an object.

Being made to penetrate someone else includes times when the victim was made to, or there was an attempt to make them, sexually penetrate someone without the victim's consent because the victim was physically forced (such as being pinned or held down, or by the use of violence) or threatened with physical harm, or when the victim was drunk, high, drugged, or passed out and unable to consent.

Sexual coercion is defined as unwanted sexual penetration that occurs after a person is pressured in a nonphysical way. In NISVS, sexual coercion refers to unwanted vaginal, oral, or anal sex after being pressured in ways that included being worn down by someone who repeatedly asked for sex or showed they were unhappy; feeling pressured by being lied to, being told promises that were untrue, having someone threaten to end a relationship or spread rumors; and pressure due to someone using their influence or authority.

Unwanted sexual contact is defined as unwanted sexual experiences involving touch but not sexual penetration, such as being kissed in a sexual way, or having sexual body parts fondled or grabbed.

Noncontact unwanted sexual experiences are those unwanted experiences that do not involve any touching or penetration, including someone exposing their sexual body parts, flashing, or masturbating in front of the victim, someone making a victim show his or her sexual body parts, someone making a victim look at or participate in sexual photos or movies, or someone harassing the victim in a public place in a way that made the victim feel unsafe.

SOURCE: Black et al. (2011, p. 17).

questions about lifetime prevalence ("How many people have ever . . ."). Respondents with an affirmative response were asked to provide "initials" to designate each offender, and follow-on questions were organized around each specific offender. In one of the follow-on questions, the respondent is asked whether the incident occurred within the past 12 months. Respondents were reminded of the date that was 12 months ago from the interview.

Moving beyond the definition used in earlier studies, the definition of rape specifically included those victimizations that occurred when the victim was "drunk, high, drugged, or passed out and unable to consent." It also measured sexual coercion by asking respondents if they experienced unwanted sexual penetration after being pressured in a nonphysical way. In addition, the survey included several items about unwanted sexual contact such as being made to penetrate someone else and several items about noncontact unwanted sexual experiences.[15]

Questions

The specific screening questions on sexual violence victimization used in the NISVS asked about victimizations experienced in respondent's lifetime and during the previous 12 months (Black et al., 2011, p. 106):

> *How many people have ever done any of the following things when you didn't want it to happen? How many people have ever...*

- *exposed their sexual body parts to you, flashed you, or masturbated in front of you?*
- *made you show your sexual body parts to them? Remember, we are only asking about things that you didn't want to happen.*
- *made you look at or participate in sexual photos or movies?*
- *harassed you while you were in a public place in a way that made you feel unsafe?*
- *kissed you in a sexual way? Remember, we are only asking about things that you didn't want to happen.*
- *fondled or grabbed your sexual body parts?*

---

[15]These noncontact unwanted sexual experiences included such things as someone exposing their sexual body parts, flashing, or masturbating in front of the victim, someone making the victim show his or her body parts, someone making the victim look at or participate in sexual photos or movies, or someone harassing the victim in a public place or in a way that made the victim feel unsafe.

When you were drunk, high, drugged, or passed out and unable to consent, how many people ever . . .

- *had vaginal sex with you? By vaginal sex, we mean that {if female: a man or boy put his penis in your vagina} {if male: a women or girl made you put your penis in her vagina}.*
- *{if male} made you perform anal sex, meaning that they made you put your penis into their anus?*
- *made you receive anal sex, meaning they put their penis into your anus?*
- *made you perform oral sex, meaning that they put their penis in your mouth or made you penetrate their vagina or anus with your mouth?*
- *made you receive oral sex, meaning that they put their mouth on your {if male: penis} {if female: vagina} or anus?*

How many people have ever used physical force or threats to physically harm you to make you . . .

- *have vaginal sex?*
- *{if male} perform anal sex?*
- *receive anal sex?*
- *make you perform oral sex?*
- *make you receive oral sex?*
- *put their fingers or an object in your {if female: vagina or} anus?*

How many people have ever used physical force or threats of physical harm to . . .

- *{if male} try to make you have vaginal sex with them, but sex did not happen?*
- *try to have {if female: vaginal} oral, or anal sex with you, but sex did not happen?*

How many people have you had vaginal, oral, or anal sex with after they pressured you by . . .

- *doing things like telling you lies, making promises about the future they knew were untrue, threatening to end your relationship, or threatening to spread rumors about you?*
- *wearing you down by repeatedly asking for sex, or showing they were unhappy?*

- *using their authority over you, for example, your boss or your teacher?*

If a respondent replied affirmatively to any of the above questions, then he or she was asked to supply "initials" to designate the specific offender. Follow-on questions were structured around the individual offender and unwanted behavior. One follow-on question asked how many times this had happened in the past 12 months:

- *Has [fill initials] [fill behavior] in the past twelve months, that is since [fill date 12 months ago]?*

Other follow-on questions were asked using similar question construction. Examples:

- *Was [fill initials] using alcohol, drugs or both the first time he/she [fill behavior]?*
- *Were you using alcohol, drugs or both the first time [fill initials] [fill behavior]? Please remember that even if someone uses alcohol or drugs, what happens to them is not their fault.*
- *Before [fill initials] had [fill (vaginal, oral, anal)] sex with you when you were drunk, high, drugged, or passed out and unable to consent, do you think you were given alcohol without your knowledge?*

Results

The results of the NISVS 2010 were summarized in *The National Intimate Partner and Sexual Violence Survey (NISVS): 2010 Summary Report* (Black et al., 2011). The NISVS Summary Report did not provide incidence rates. It provided prevalence rates. Table 5-4 shows the survey's 2010 prevalence estimates for completed forced penetration, attempted forced penetration, and alcohol- or drug-facilitated completed penetration—both 12-month and lifetime—by gender. Unfortunately, the 12-month prevalence estimates for rape victimizations for males did not meet the reliability criteria[16] and were, therefore, not reported. The 12-month prevalence rates for all adults also were not published.

[16]Statistical thresholds applied to all estimates in the 2010 Summary Report.

TABLE 5-4 12-Month and Lifetime Prevalence of Rape Victimization, National Intimate Partner and Sexual Violence Survey, 2010

	12-Month Prevalence		Lifetime Prevalence	
Type of Victimization	Weighted Percentage of U.S. Adult Women	Estimated Number of Victims[a]	Weighted Percentage of U.S. Adult Women and Men	Estimated Number of Victims
Women, 18+				
Completed forced penetration	0.5	620,000	12.3	14,617,000
Attempted forced penetration	0.4	519,000	5.2	6,199,000
Completed alcohol/drug-facilitated penetration	0.7	781,000	8.0	9,524,000
Total completed, attempted, and completed alcohol/drug-facilitated penetration[b]	1.1	1,270,000	18.3	21,840,000
Men, 18+				
Completed forced penetration	[c]	[c]	0.9	970,000
Attempted forced penetration	[c]	[c]	0.4	499,000
Completed alcohol/drug-facilitated penetration	[c]	[c]	0.6	685,000
Total completed, attempted, and completed alcohol/drug-facilitated penetration[b]	[c]	[c]	1.4	1,581,000

[a]Rounded to nearest thousand.

[b]Respondents who reported more than one subcategory are counted only once in the total estimate but are included in each relevant subcategory. For example, victims of completed forced penetration and alcohol- or drug-facilitated penetration are included in each of these subtypes of rape but are counted only once in the total estimate of rape prevalence.

[c]Estimate not reported. The relative standard error of the estimate is greater than 30 percent, or the cell size is 20 or less.

6

Comparison of Rape and Sexual Assault Across Data Sources

The sources of data on sexual victimizations discussed in this report have different foci, use different methodologies, and provide different results. At this point, definitive conclusions regarding which data source produced the most accurate estimates of rape and sexual assault would be useful. However, the panel acknowledges that it cannot scientifically make such conclusions in this report. The first barrier to such conclusions: the panel focused on the National Crime Victimization Survey (NCVS) and not the other sources. It examined the NCVS under the structure of total survey error (Chapters 7, 8, and 9), and these analyses allowed the panel to draw conclusions and make recommendations regarding the NCVS. The panel did not provide the same deliberative focus on each of the other sources of data, in part because of limited time and resources. Thus we know a good deal about the potential errors in the NCVS and much less about the potential errors in the other sources. Specifically, this does not mean that the other sources have fewer potential errors, only that these errors are not analyzed in this report. A second barrier: the target populations and definitional constructs (of what was being measured) are inconsistent across sources. It is a case of comparing apples and oranges. Third, only the NCVS and the Uniform Crime Reports (UCR) provided estimates over time, with the other sources providing estimates only for specific (and different) points in time. Thus, the complexity of different concepts, measurement approaches, and timing made definitive comparisons very problematic, and the panel did not have the time and resources available to attempt such a task.

With that said, a better understanding of the differences between these sources and their measurement approaches can lead to improvements in the measurement of rape and sexual assault on Bureau of Justice Statistics (BJS) surveys. Therefore, this chapter, drawing on Appendix D and Bachman (2012), summarizes and highlights what the panel learned from the comparisons among the five surveys and one administrative source covered in this report.

1. UCR summary system (ongoing),
2. NCVS (ongoing),
3. National Women's Study (NWS) (1989-1990),
4. National Violence Against Women Study (NVAWS) (1995),
5. National College Women Sexual Victimization Study (NCWSV) (1997), and
6. National Intimate Partner and Sexual Violence Survey (NISVS) (2010, and perhaps ongoing).

Our comparisons are discussed in terms of five factors: the definitions used for rape and sexual assault; context in which data are collected; target population, sampling frame, and sample size; data collection mode, response rates, and adjustments for nonresponse; and the resulting measures of victimization.

DEFINITIONS USED

The definitions used for rape vary, sometimes substantially, among the six data sources (see Table 6-1). The table also shows whether the source collected information on attempted rape and other forms of sexual assaults as well as rape.

The UCR definition (used through 2012) is clearly the most restrictive. It restricts rape counts to male on female attacks with penile-vagina penetration. Attempted rapes are counted, but all other forms of sexual victimizations are included in a general "assault" category. The revised definition, scheduled for implementation in 2013, will provide a broader base for reports of rape and attempted rape. This change should result in a larger number of crimes being counted as rape and fewer crimes being counted in the "assault" category. Importantly, the UCR only measures incidents reported to police. This is an important difference with the other data sources, and the new Federal Bureau of Investigation (FBI) definition will not change this difference.

The NCVS has a broader definition of rape. It includes male and female victims and offenders. It includes penetration (vaginal, anal, and oral) by penis, other body parts, and other objects. It also separately measures attempted rape and a fairly wide range of sexual assaults, including verbal

TABLE 6-1 Definitions of Rape and Sexual Assault, by Data Source

Data Source	Basic Description	Estimates
Uniform Crime Reports (UCR) Summary Reporting System	**Rape** includes only male offenders and female victims, with penile penetration of a vagina. **Attempted rape** is counted separately. Other forms of **sexual assault** are included in a general category of assault and not summarized separately or with rape. An updated definition has been developed and was scheduled to be used beginning January 2013. The updated definition covers male and female victims and penetration with other (than penis) body parts and objects. It covers anal penetration and oral penetration by a sex organ. The definitional change does not change sexual assault, which is still included in the general category of assault.	Rape and attempted rape. Does not include other forms of sexual assault.
National Crime Victimization Survey (NCVS)	**Rape** includes psychological coercion as well as physical force. Forced sexual intercourse means vaginal, anal, or oral penetration by the offender(s). It also includes incidents in which the penetration is by a foreign object. It includes attempted rapes, male as well as female victims, and both heterosexual and homosexual rape. **Sexual assault** included in this category includes a wide range of victimizations that are separate from rape or attempted rape. These crimes include attacks or attempted attacks generally involving unwanted sexual contact between victim and offender. Sexual assaults may or may not involve force and include such behavior as grabbing or fondling. Sexual assault also includes verbal threats.	Measures rape, attempted rape, and a wide category of sexual assault.
National Women's Study (NWS)	**Rape** was defined as an event that occurred without the woman's consent, involved use of force or threat of force, and involved sexual penetration of victim's vagina, mouth, or rectum. The NWS results included only female victims and measured prevalence rather than the number of incidents.	Measures rape. Does not measure attempted rape or other forms of sexual assault.

continued

TABLE 6-1 Continued

Data Source	Basic Description	Estimates
National Violence Against Women Study (NVAWS)	**Rape** was defined as an event that occurred without the victim's consent and that involved the use or threat of force to penetrate the victim's vagina or anus by penis, tongue, fingers, or object, or the victim's mouth by penis. The definition **included both attempted and completed rape.**	Measures rape and attempted rape. Does not measure other forms of sexual assault.
National College Women Sexual Victimization Study (NCWSV)	**Rape** is unwanted completed penetration by physical force or the threat of physical force. Penetration includes penile-vaginal, mouth on the victim's genitals, mouth on someone else's genitals, penile-anal, digital-vaginal, digital-anal, object-vaginal, and object-anal.	Measures rape and attempted rape, as well as various forms of sexual assault.
	Attempted rape is the unwanted attempted penetration by force or the threat of force.	
	Threat of rape is the threat of unwanted penetration with force and threat of force. The NCWSV results include only female college students as victims.	
National Intimate Partner and Sexual Violence Survey (NISVS)	**Rape** is defined as any completed or attempted unwanted vaginal (for women), oral, or anal penetration through the use of physical force (such as being pinned or held down, or by the use of violence) or threats to physically harm. It includes times when the victim was drunk, high, drugged, or passed out and unable to consent. Rape is separated into three types: completed forced penetration, attempted forced penetration, and completed alcohol- or drug-facilitated penetration.	Measures rape and attempted rape, as well as various forms of sexual assault.
	Among women, rape includes vaginal, oral, or anal penetration by a male using his penis. It also includes vaginal or anal penetration by a male or female using his or her fingers or an object.	
	Among men, rape includes oral or anal penetration by a male using his penis. It also includes anal penetration by a male or female using his or her fingers or an object.	

SOURCES: Data from Black et al. (2011); Bureau of Justice Statistics (n.d.-b); Federal Bureau of Investigation (2004); Fisher, Cullen, and Turner (2000); Kilpatrick, Edmunds, and Seymour (1992); Tjaden and Thoennes (2000).

threats. It reports separate estimates of rape and sexual assault, and also reports the two categories together. When comparing data from the NCVS with other sources, one has to decide which of the NCVS's numbers to use.

The NWS used the term "forcible rape." It included much of the broader set of penetration victimizations by force or threat of force included in the NCVS definition of rape. The study's estimates did not include attempted rape or other forms of sexual assault. The estimates it produced were more narrowly focused than the NCVS because its target population was only adult women.

The NVAWS measured victimizations of both male and female adults. The definition of rape included penetration and attempted penetration (vaginal, oral, and anal) by force or threat of force. It did not cover other forms of sexual assault.

The NCWSV targeted only college women. Thus, sexual victimizations against men and against women not in college were not included. The study measured 12 different types of victimization (see Table 5-2 in Chapter 5). Completed rape included penetration (vaginal, oral, and anal) by force or threat of force.

The NISVS measures both completed and attempted rape as defined by penetration with use or threat of physical force. It attempts to measure victimizations of both adult males and females. However, because of limited sample size in the first (2010) survey, estimates were published only for females. It extends the definition of rape to include penetration when the victim was unable to consent by being drunk, high, drugged, or passed out.

The panel next compared the above definitions with the commonalities of legal definitions we found across jurisdictions and presented in Chapter 2. Table 6-2 provides a summary of this comparison. Across the sources, there was less uniformity among the data sources regarding the inclusion of nonpenetration sexual assault and in gender restriction. The NISVS was the only source that specifically included alcohol- and drug-facilitated penetration as part of forced sexual activities. The panel identified this as a missing component to the NCVS definition (see Recommendation 10-7).

SURVEY CONTEXT

The context of a survey is very important to both response rates and to the quality of responses that are received. A simple change in context can make a big difference. For example, when the National Survey of Drug Abuse changed its name to the National Survey of Drug Use and Health, reported drug use increased (Office of Applied Studies, 2003). Context can be established in a number of ways, including the prior questions in a questionnaire (Holyk, 2008, p. 42):

> The term context effect refers to a process in which prior questions affect responses to later questions in surveys. Any survey that contains multiple questions is susceptible to context effects. Context effects have the potential to bias the thinking and answers of survey respondents, which reduces the accuracy of answers and increases the error in survey measurement.

In comparing the overall context of data sources for measurements of rape and sexual assault, two sources focus specifically on crimes: the UCR and the NCVS. The UCR summarizes "crimes known to police"—those that are both reported to and recorded by police. The NCVS's goal is to measure the victimization rate by type of crime. The NCVS is a national crime survey, and the questionnaire asks many questions about different types of crimes as well as the well-being of victims.

The other four sources (NWS, NVAWS, NCWSV, and NISVS) do not frame the survey around/on criminal victimization. They focus instead on the situations in which the respondent may have experienced nonconsensual or unwanted sexual contact. These surveys also collect additional information about the respondent's well-being.

The panel believes that survey context is likely a major contributor of differences in the estimates of rape and sexual assault between the several sources.

TARGET POPULATION, SAMPLING FRAMES, AND SAMPLE SIZE

Target Populations and Sampling Frames

The target populations for the six surveys are different, with resulting effects on the estimates. The NWS targeted adult (18+ years) women and made no estimates for men. The NCWSV targeted a narrower group of women, only those attending college. The NVAWS and the NISVS both targeted adult (18+ years) men and women. However, the NISVS only published estimates (12-month prevalence) of rape and attempted rape for women for its first implementation (in 2010). The NCVS targets both men and women with a broader age range (12+ years).

The sampling frames were different for different sources. Three surveys (NWS, NVAWS, and NISVS) are based on geographically spread random digit dialing (RDD) frames. The RDD frames cover only U.S. households that have telephones. It is important to consider whether this undercoverage is serious. The NWS and the NVAWS were conducted between 1989 and 1995. The 1990 census showed that 5.2 percent of U.S. households had no telephones. The percentage was above 10 percent in five states.[1] The next

[1]Arkansas, 10.9 percent; Kentucky, 10.2 percent; Mississippi, 12.6 percent; New Mexico, 12.4 percent; West Virginia, 10.3 percent.

TABLE 6-2 Comparing Definitions Used for Rape with Commonalities Found Across Jurisdictional Legal Codes

Commonalities	Comparison with Definitions Used by Data Sources
Victimization not restricted by gender. Both male and female victims and offenders.	The UCR definition is currently restricted by gender. The updated definition is not. The NCVS is not restricted by gender. The NWS and the NCWSV targeted only women victims but did not restrict the gender of the offender. The NVAWS and the NISVS were not gender restricted.
Rape involves a broad range of penetrations.	The UCR definition is currently restricted to penile penetration of a vagina. The updated definition is not. The NCVS, NWS, NCWSV, NVAWS and NISVS include a broad range of penetrations.
Purpose is for sexual arousal or degradation.	This is not a specifically stated component of the definitions of any of the sources but probably does not need to be. In legal statutes, this is used to distinguish assaults from such things as medical exams.
Use of force or threat of force against the victim or another person.	Consistent for all sources.
Lack of consent or lack of capacity to consent.	Lack of consent is consistent for all sources. The NISVS specifically includes questions related to the lack of capacity to consent due to alcohol and/or drug use.
Sexual assault includes a fairly wide range of victimizations that involve unwanted non-penetration sexual contact.	The UCR includes sexual assault with other types of assault and does not have a separate category. The NCVS has a definition of sexual assault that is consistent with the commonality. It summarizes it separately and also combines it with rape and attempted rape. The NWS does not measure attempted rape or sexual assault. The NVAWS includes attempted rape but not other types of sexual assault. The NCWSV and the NISVS have definitions of sexual assault consistent with this commonality.

NOTES: Commonalities described in Chapter 2. NCVS = National Crime Victimization Survey, NCWSV = National College Women Sexual Victimization Study, NISVS = National Intimate Partner and Sexual Violence Survey, NVAWS = National Violence Against Women Study, NWS = National Women's Study, UCR = Uniform Crime Reports.
SOURCES: Data from Black et al. (2011); Bureau of Justice Statistics (n.d.-b); Federal Bureau of Investigation (2004); Fisher, Cullen, and Turner (2000); Kilpatrick, Edmunds, and Seymour (1992); Tjaden and Thoennes (2000); Tracy et al. (2012).

census, in 2000, showed the percentage of households without telephone was only 2.4 percent of households, but the same five states still lagged in coverage (U.S. Census Bureau, 2000).[2] Thus, there is some concern about coverage in these two surveys for poorer households (the ones that generally lack telephones) in those states, but the potential coverage error is small enough that it was unlikely to have much effect on the estimates.

In more recent years, there has been an increase in the percentage of cell-phone-only households (with cell phones but no landline), from approximately 3 percent in 2003 to 34 percent in 2012 (Blumburg and Luke, 2012). The overall trend to cell phones instead of landlines is significant (Hall, Carlson, and CyBulski, 2011, p. 2):

> This drastic change in cell phone usage has significantly affected the coverage of surveys that use random digit dialing sampling. Because of this trend in cell phone usage over the last decade, using only a landline-based RDD sample results in reducing the coverage of the population.

Responding to these changes, the NISVS augmented its RDD sample of landline phone numbers to include a cell phone sample. In 2011, only about 2 percent of U.S. households had no telephone, landline, or cell phone (Blumburg and Luke, 2012).

Two of the studies reviewed by the panel relied on cluster sampling. Sampling for the NCWSV involved two stages of selection. The first stage frame was a list of academic institutions stratified by total student enrollment and institution location, and the second stage was a list, for the selected institutions, of women enrolled in the fall of 1996.

The sample design for the NCVS begins with a selection of primary sampling units (PSUs) from the Census Bureau, and then uses the Master Address File supplemented with the New Building Permits frame and the Group Quarters frame (see Chapter 4 for more details).

The UCR data come from an administrative source (voluntary reports from law enforcement agencies); they are not based on a random sample.

Sample Size

Table 6-3 displays information on sampling frames and sample sizes. The sample size for the NCVS is substantially larger than for the other surveys. Its sample size has fluctuated with annual budget changes over the past years: the smallest number of interviews was 134,041 in 2005, and the largest number was 181,205 in 1994.

The NWS interviewed 4,008 women, approximately 4 percent of the

[2] Arkansas, 5.4 percent; Kentucky, 4.7 percent; Mississippi, 6.5 percent; New Mexico, 5.7 percent; West Virginia, 4.7 percent.

people interviewed in the 1990 NCVS (95,000 interviews). The sample size for the NCWSV was approximately the same as the NWS, with 4,446 interviews. However, the NCWSV was targeting a much smaller population. The sample size for the NVAWS was 16,000 interviews, and the sample size for the NISVS was 18,049 interviews. Thus, the sample size for these two surveys were only about 11 percent of the sample size for the 2011 NCVS. Smaller sample sizes are a particular problem when measuring a low incidence event, such as rape. For example, the NVAWS, with a sample size of about 16,000, found only 24 women and 8 men who reported having been raped. And the NISVS was unable to publish estimates of male victimizations because of the small number of reported victimizations.

The UCR is essentially a census of all police reports from approximately 18,000 participating jurisdictions, which covers approximately 90 percent of all jurisdictions. It has no incident-level or individual-level records. Coverage in metropolitan areas is slightly higher than in rural areas. These police reports are widely believed to be missing a substantial percentage of the rapes and sexual assaults that occur, as much as 65-80 percent.

DATA COLLECTION MODE AND RESPONSE RATES

Data collections for all five surveys are interviewer administered and rely heavily on telephone interviewing. Beyond that generality, however, there are differences (see Table 6-3). The NCVS is an ongoing panel survey, with selected households in the survey for 3 years. This is different from the other surveys. The NCVS begins with a presurvey letter and an in-person visit for wave 1. It uses telephone interviews, conducted by the field representatives, for other waves if feasible (see Chapter 4). The NCWSV also began with a presurvey letter so that the telephone interview that followed was not based on a cold contact. The others (NWS, NVAWS, and NISVS) were all RDD surveys with a cold initial contact from a centralized telephone facility.[3] All interviews used computer-assisted interviewing technology. This discussion is not relevant for the UCR, which uses administrative data.

Nonresponse can affect both survey estimates and their estimated variances. For a survey, nonresponse bias is dependent on both the size of the nonresponse and to the extent differences exist between respondents and nonrespondents regarding important variables being measured in the survey. Because of these potential effects, response rate has been one indicator used to assess survey quality.

The rates for the data sources discussed in this chapter varied consider-

[3]The NISVS was able to match 50 percent of its landline sample with addresses. These households were sent an advance letter.

TABLE 6-3 Sampling Information, Data Collection Mode and Response Rates, by Data Source

Data Source	Year	Sampling Frame
Uniform Crime Reports (UCR) Summary Reporting System	Since 1929	Police reports
National Crime Victimization Survey (NCVS)	Annually since 1991 Predecessor survey since 1972	2-stage cluster sample, with area-based PSUs. Addresses sampled from the Census Bureau's Master Address File, the new building permits frame, and the group quarters frame Target—males and females 12+ years
National Women's Study (NWS)	1989-1990	RDD: 2-stage sample, with primary stage geographic areas within the United States; 2nd stage a random digit dialing using landline frame within selected geographic areas
National Violence Against Women Study (NVAWS)	Nov 1995 through May 1996	RDD: National random digit dialing (landline) sample, selected within census regions
National College Women Sexual Victimization Study (NCWSV)	1996-1997	2-stage sample, with primary stage stratified list of 2- and 4-year colleges; 2nd stage a sample of women enrolled in fall 1996
National Intimate Partner and Sexual Violence Survey (NISVS)	2010	RDD: Random digit dialing using both landline and cell phone frames

[a]Sample size—individuals interviewed.

[b]Response rate calculated based on AAPOR Standard 4.

[c]The participation rate for the NVAWS was calculated by dividing the number of completed interviews (including those that were screened out because they were ineligible) by the total number of completed interviews, screened-out interviews, refusals, and terminated interviews (Tjaden and Thoennes, 2000, p. 4).

Sample Size[a]	Data Collection Mode	Response Rates
100% from participating 18,000 jurisdictions	Administrative records	NA
n = 143,122 individuals interviewed, male and female, 12+ years, from 79,802 households (2011)	Interviewer administered. In-person interview on first and last waves. Telephone interview on other waves, to the extent feasible	2011—88% for individual respondents
n = 4,008, female only, 18+ years	Interviewer administered with cold telephone contact on wave 1, with subsequent waves also administered with telephone interview	34% response rate[b] 85% participation rate
n = 16,000, male and female, 18+ years	Interviewer administered with cold telephone contact	34% response rate[b] 72% participation rate for females and 69% participation rate for males[c]
n = 233 institutions n = 4,446 female college students	Interviewer administered. Presurvey letter followed by telephone interview	67% response rate,[b] with 86% participation rate
n = 18,049 (2010), male and female, 18+ years	Interviewer administered with cold telephone contact	2010—34% response rate

NOTE: AAPOR = American Association for Public Opinion Research, PSUs = primary sampling units.
SOURCES: Data from Black et al. (2011); Bureau of Justice Statistics (n.d.-b); Federal Bureau of Investigation (2004); Fisher, Cullen, and Turner (2000); Kilpatrick, Edmunds, and Seymour (1992); Tjaden and Thoennes (2000).

ably. The NCVS has both the largest sample size and the highest response rate, 88 percent. The NCWSV achieved a modest response rate of 67 percent. RDD surveys invariably have lower response rates because of "screen outs" needed to obtain a qualified respondent and the effect of cold calling. The three RDD surveys (NWS, NVAWS, and NISVS) had similar response rates—NWS, 34 percent; NVAWS, 34 percent; NISVS, 34 percent—all substantially lower than that achieved by the NCVS and the NCWSV. Cooperation rates (which are calculated once a qualified respondent is reached) were higher in each of these three surveys.

MEASURES OF RAPE AND SEXUAL ASSAULT

The number of rapes and sexual assaults from the four surveys and the UCR are shown in Figure 6-1.[4] One only has to look at the graphic to understand that these data sources are both measuring different things and measuring things differently. Table 6-4 shows the estimates of number of rapes from all six sources. Table 6-5 shows estimates for females aged 18+ only. The lowest 12-month estimate for rape is measured by the UCR. This is not surprising because the UCR has the most restrictive definition of rape and only measures rapes (and attempted rapes) that are known to the police. Therefore, the UCR is assembled in ways that make it vulnerable to major undercounting.

The estimates from the NISVS are the largest. It is important to note that the NISVS published 12-month prevalence numbers and only for women: if the 12-month estimates were of incidents and for all adults, then the numbers would be even larger. The number of females raped or sexually assaulted (adult females only) estimated by the NISVS is 5 times larger than the number of incidents measured by the NCVS (including series victimization) for rape and sexual assault for females (age 12+) in 2010, twice as large as the prevalence number estimated by the NWS (adult females, completed rape only in 1990),[5] and 30 percent greater than measured by the NVAWS (both male and female adults, completed and attempted rape but no other forms of sexual assault in 1995). This differential between the NISVS and the other surveys is surprising. The definitions are not identical, but they are roughly consistent. The NISVS, along with the NWS and NVAWS, used RDD survey designs. All three of these surveys had the same response rate, 34 percent.

The panel attempted to look at confidence intervals for comparisons

---

[4]The rates from the NCWSV are excluded because the survey covered only a small at-risk population—college women.

[5]However, the NISVS prevalence estimate for completed rape only (620,000—see Table 5-4) is close to the NWS estimate of incidents of completed rape for adult women (683,000).

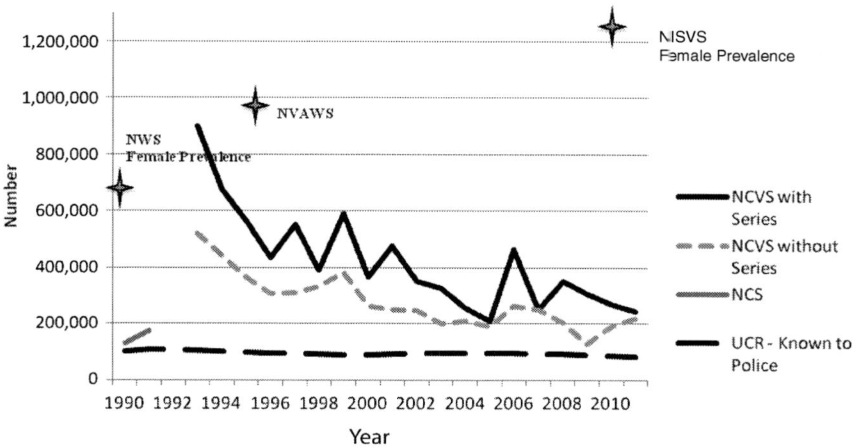

FIGURE 6-1 Number of rapes, by data source.
NOTES:

- NWS and NISVS estimates are for adult (18+) women only.
- NWS and NISVS estimates are prevalence rates and not incidence rates.
- NWS estimates do not include attempted rape or other sexual assaults.
- NCVS estimates are for rape and sexual assault, ages 12+ years.
- UCR estimates are for rape and attempted rape that are known to police (no age limit).

Estimates from the NCWSV are specifically for college women and thus not comparable to the others and not included in this figure. NCS = National Crime Survey, NCVS = National Crime Victimization Survey, NCWSV = National College Women Sexual Victimization Study, NISVS = National Intimate Partner and Sexual Violence Study, NVAWS = National Violence Against Women Study, NWS = National Women's Study, UCR = Uniform Crime Reports.
SOURCES: Panel-developed graphic using data from Black et al. (2011); Bureau of Justice Statistics (1991, 1992, 1993, 1994, 1995, 1996, 1997, 1998, 1999, 2000, 2001, 2002a, 2003, 2004, 2005, 2006, 2007, 2008a, 2009, 2010, 2011, 2012a, n.d.-a); Federal Bureau of Investigation (n.d.-c); Fisher, Cullen, and Turner (2000); Kilpatrick, Edmunds and Seymour (1992); and Tjaden and Thoennes (2000).

as well as the point estimates described above. This attempt was somewhat unsatisfying because we were unable to obtain standard errors from the NWS or the NISVS.[6] Rand and Rennison (2005) compared the NVAWS with the NCVS, and we extracted the NVAWS standard errors from that journal article. For 1995, the NCVS estimate of the number of rapes and sexual assaults for adult women was 403,735, including series victimization

[6]The NISVS published standard errors for their lifetime prevalence rates but not for the 12-month prevalence rates, which were needed for our comparisons.

TABLE 6-4 Estimates of Number of Rapes and Sexual Assault by Year, by Source

Year	Gender of Victim	Source					
			NCVS				
		UCR	No Series[a]	Series[b]	NWS	NVAWS	NISVS
1990	F[c]	102,555	106,000	NA	683,000[d]		
	All		130,260	NA			
1991	All	106,593	174,000	NA			
1992	All	109,062	NA	NA			
1993	All	106,014	521,223	898,239			
1994	All	102,216	443,509	674,291			
1995	All	97,470	363,527	563,249		987,362	
1996	All	96,252	307,100	437,198			
1997	All	96,153	311,110	553,523			
1998	All	93,144	332,500	391,101			
1999	All	89,411	383,170	591,460			
2000	All	90,178	260,950	366,747			
2001	All	90,863	248,250	476,578			
2002	All	95,235	247,730	349,805			
2003	All	93,883	198,850	325,311			
2004	All	95,089	209,880	255,769			
2005	All	94,347	190,592	207,760			
2006	All	92,757	260,940	463,598			
2007	All	90,427	248,277	248,277			
2008	All	90,479	203,830	349,691			
2009	All	89,241	125,910	305,574			
2010	F[c]			253,555			1,270,000[e]
	All	84,767	188,380	268,574			
2011	All			243,803			

NOTES: NCVS = National Crime Victimization Survey, NISVS = National Intimate Partner and Sexual Violence Survey, NVAWS = National Violence Against Women Study, NWS = National Women's Study, UCR = Uniform Crime Reports.

[a]Series victimizations are excluded.

[b]Includes series victimizations up to 10 incidents.

[c]Female adults.

[d]Prevalence estimate for completed rapes only; does not include attempted rape or other forms of sexual assault.

[e]Prevalence estimate (number of victims) rather than the number of victimizations.

SOURCES: Panel-developed table using data from Black et al. (2011, 2012a, n.d.-a); Bureau of Justice Statistics (1991, 1992, 1993, 1994, 1995, 1996, 1997, 1998, 1999, 2000, 2001, 2002a, 2003, 2004, 2005, 2006, 2007, 2008a, 2009, 2010, 2011, 2012a, n.d.-a); Federal Bureau of Investigation (n.d.-c); Fisher, Cullen, and Turner (2000); Kilpatrick, Edmunds and Seymour (1992); and Tjaden and Thoennes (2000).

TABLE 6-5 Estimates of Number of Rapes and Attempted Rapes of Females, 18 Years of Age and Older for Selected Years, by Source

	Source					
		NCVS[b]				
Year	UCR[a]	No Series	Series	NWS[c]	NVAWS[d]	NISVS[e]
1990	102,555	106,000	NA	683,000		
1995	97,470	NA	403,735		876,064	
2010	85,593	NA	209,740			1,270,000

NOTES: These numbers represent the most consistent comparison that the panel was able to identify. Even so, they have differences, which are listed below. NCVS = National Crime Victimization Survey, NISVS = National Intimate Partner and Sexual Violence Survey, NVAWS = National Violence Against Women Study, NWS = National Women's Study, UCR = Uniform Crime Reports.

[a]UCR Forcible rape and attempted rapes known to police. Includes females, all ages.

[b]NCVS Rape, attempted rape, and sexual assault, for 1990 only, includes females ages 12+ years. For 1996 and 2010, includes females, 18+ years.

[c]NWS Forcible rape only. Does not include attempted rape. Count of victims rather than victimizations.

[d]NVWS Forcible rape and attempted rape.

[e]NISVS Rape and attempted rape. Count of victims rather than victimizations.

SOURCES: Data from Black et al. (2011); Bureau of Justice Statistics (1991, 1996, 2011, n.d.-a); Federal Bureau of Investigation (n.d.-c); Kilpatrick, Edmunds, and Seymour (1992); Tjaden and Thoennes (2000).

(see Table 6-5). The standard error was 41,643.[7] The 95 percent confidence interval was [322,115, 485,355]. The NVAWS estimate of the number of forcible rapes and attempted rapes was 876,064. The NVAWS had a much smaller sample size than did the NCVS, and as expected, its standard error was much larger: 467,098. The 95 percent confidence interval for the NVAWS estimate was [–39,448, 1,791,576]. Thus the 95 percent confidence interval for the NCVS was contained completely within the 95 percent confidence interval for the NVAWS. Rand and Rennison (2005) made a similar finding, with no statistical difference in those estimates.

In reviewing all of this material, the panel thinks that it is highly likely that the NCVS is underestimating rape and sexual assault. The panel, with limited resources, was not able to measure the extent of such an undercount, but the pattern is one that shows lower estimates of rape and sexual

[7]This is the standard error for estimate of rape and sexual assault for all women. The panel did not have the actual standard error for women 18 years of age and older.

assault in the NCVS than the estimates published from the other surveys.[8] Thus, the panel looked in more detail at the error profile of the NCVS to better understand procedures that might be contributing to this undercount. This analysis is contained in Chapters 7, 8, and 9.

[8]This was the case even though these other surveys were often more restrictive in what they measured, such as focusing on adults only, or women only, or not including attempted rape or other forms of sexual assault.

7

Potential Sources of Error in the NCVS: Sampling, Frame, and Processing

The nation needs accurate measurements of victimization rates to allocate resources to fight crime, support victims' needs, and shape policies and programs to deter these crimes in the future. The National Crime Victimization Survey (NCVS), which is administered by the Bureau of Justice Statistics (BJS), is currently the major tool available to measure these rates and victim characteristics. As discussed in the preceding chapters, there is controversy as to whether the incidence of rape and sexual assault is being underestimated on the NCVS, in part because other sources of data have shown higher levels of victimization than estimated through the NCVS. These differences reflect, in part, the clear definitional differences and methodological differences among the sources, which in turn affect the estimated victimization levels.

The panel could not ascertain which data source provided the most accurate estimates of rape and sexual assault. Even though the other sources (excluding the Uniform Crime Reports [UCR]) showed larger estimates than did the NCVS (or National Crime Survey), the panel is not concluding that "bigger is better." With that said, the higher rates estimated by the several reviewed surveys lend support to concerns about a potential underestimate by the NCVS. These concerns, as well as the original charge to the panel (see Box 1-1 in Chapter 1), led to the panel's close analysis of the NCVS. It is important to note that the panel's work focused on the NCVS and did not examine as closely the other sources of data on rape and sexual assault described in Chapter 5. By addressing only the NCVS in this and the next three chapters, the panel is not implying that there are more issues with the NCVS than with the others.[1]

To assess potential issues with the survey, including reasons for this possible underestimate, the panel examined the NCVS using the structure of total survey error. Total error involves a holistic view of all potential errors in a survey program, including both sampling error and nonsampling error. Biemer (2010), in the *Handbook of Survey Research*, outlines the components of nonsampling error in surveys: specification error, frame error, nonresponse error, measurement error, and processing error. *Specification error* arises when the construct underlying an observed variable, y, differs from the desired construct, x—that is, the construct that data analysts and other users prefer. *Frame error* arises in the process of constructing, maintaining, and using the sampling frame(s) for selecting the survey sample. It includes the inclusion of nonpopulation members (overcoverage), exclusion of population members (undercoverage), and the duplication of population members. Frame error also includes errors in the auxiliary variables associated with the frame units (sometimes referred to as content error) as well as missing values for these variables. *Nonresponse error* encompasses both unit and item nonresponse. Unit nonresponse occurs when a sampled unit does not respond to any part of a questionnaire. Item nonresponse occurs when the questionnaire is only partially completed because an interview was prematurely terminated or some items that should have been answered were skipped or left blank. *Measurement error* includes errors arising from respondents, interviewers, survey questions and factors that affect survey responses. Data *processing error* includes errors in editing, data entry, coding, computation of weights, and the tabulation of the survey data. It also includes errors arising from fitting models for various purposes such as imputation, derivation of new variables, disclosure avoidance and so forth.

This chapter reviews potential sampling problems for the NCVS in measuring rape and sexual assault and then reviews potential problems for two nonsampling errors: frame and processing. Chapter 8 reviews the other three nonsampling errors: nonresponse, specification, and measurement.

SAMPLING ERROR

Sampling error occurs because survey information is observed from only a sample of the target population instead of from the entire population. In general, increasing the size of the sample decreases sampling error.

[1]In fact, the panel was somewhat discomfited by the estimates from the National Intimate Partner and Sexual Violence Survey (NISVS), which are so much larger than estimates from other sources of data. The field would benefit from a rigorous error evaluation of that survey to better understand those differences.

Because the costs of increasing sample size are not trivial, survey practitioners strive to create operationally efficient sample designs that can provide sufficient coverage of the target population while keeping both sample size and sampling error within specified bounds.

The target population of the NCVS is the noninstitutionalized population of the United States, 12 years of age and older. This includes "residents living throughout the United States, including persons living in group quarters, such as dormitories, rooming houses, and religious group dwellings" (Bureau of Justice Statistics, 2008b). As detailed in Chapter 4, the Census Bureau uses a multistage sample design for the NCVS, with primary sampling units (PSUs) coming from an area frame and secondary units of addresses selected within the sampled PSUs. The household at a selected address is contacted in person by a Census Bureau field representative, and each household member, 12 years of age and older, is identified for a separate interview. Subsequent waves of interviewing are conducted by the field representative, mostly by telephone.

The NCVS features a rotating panel design, with selected addresses included in the sample for seven waves of collection over 3 years. Every 6 months, a new rotation group is rotated into the NCVS and an existing group is rotated out. The second stage sample is of addresses, so if individuals move into or out of the housing unit during the time that their address is in the NCVS sample, then field interviewers will contact current residents at the originally sampled address.

BJS derives annual estimates of victimization levels and rates by accumulating data from all rotation groups across all data collections in the year. The sampling weights used in these estimates are adjusted for nonresponse and calibrated to known population totals. There is also an adjustment for data collected during a first interview that compensates for potential telescoping (see details in Chapter 4).

Large Coefficients of Variation

The coefficient of variation (CV), which is defined as the standard error of a survey estimate divided by the estimate itself (expressed as a percent), provides a relative measure of the sampling error associated with survey estimates. Table 7-1 shows national-level estimates and their CVs from the NCVS for rape and sexual assault and serious violent crimes for 3 years: 2002, 2010, and 2011. For the category of all serious violent crimes (which includes rape and sexual assault, robbery, and aggravated assault), the sampling error at the national level for both the number of victimizations and the victimization rate is approximately 6 percent and appears to be fairly stable from year to year. However, for rape and sexual assault (as measured in 2011), which account for only 13 percent of all serious vio-

TABLE 7-1 NCVS Estimates and Coefficients of Variation (CVs): Rape and Sexual Assault and Serious Violent Crimes (including series victimization)

		Number of Victimizations		Rate per 1,000 People (12+ years and older)	
Year	Category of Victimization	Estimate	CVs (%)	Estimate	CVs (%)
2002	Rape and sexual assault	349,810	10.9	1.5	13.3
	All serious violent crimes	2,306,710	5.2	10.0	5.0
2010	Rape and sexual assault	268,570	13.4	1.0	10.0
	All serious violent crimes	1,694,840	6.4	6.6	6.1
2011	Rape and sexual assault	243,800	14.3	0.9	11.1
	All serious violent crim	1,852,650	6.2	7.2	5.6

SOURCE: Data from *Criminal Victimization, 2011* (Bureau of Justice Statistics, 2012a).

lent crime victimizations, the CVs for the number of those victimizations are approximately 14 percent at the national level with more year-to-year variation. The CVs for victimization rates for rape and sexual assault are slightly smaller: 10–11 percent in the two most recent years.

As important as national-level estimates of rape and sexual assault are, there is an equivalent need for quality estimates for certain subpopulations to ascertain which demographic groups are more "at risk" to become victims and to look at regional differences in criminal victimization levels. These more focused estimates are important because they allow for better allocation of resources to prevent crime and support victims.

Unfortunately, the sampling error for estimates of victimization rates for many subpopulations of interest can become quite large on the NCVS because there are very few affirmative responses to questions about serious violent criminal victimization in the sampled groups. Thus, BJS does not provide estimates for rape and sexual assault for these subpopulations; they only provide estimates for the larger category, serious violent crimes.

For the aggregated category, serious violent crime, Table 7-2 shows that the CVs at the national level are approximately 6 percent. However, the CVs for important subpopulations are much higher because of their smaller sample sizes. For example, the NCVS estimates that blacks experienced an estimated serious violent victimization rate of 10.8 percent in 2011, which were 65 percent higher than that experienced by whites, 6.5 percent, and the CVs for blacks were high (13 percent). American Indians/Alaska Natives experienced an estimated serious violent victimization rate of 47.3 percent in 2010 and 12.6 percent in 2011, and the CVs for those years were 24 and 51 percent, respectively. It is clear that the sampling errors for these important "at-risk" subpopulation were large and the estimates were very unstable from year to year.

Age groups show different estimated levels of victimizations, with highest rates for people 24 years of age and younger. The CVs for all age groups were above 10 percent in 2010 and 2011. The estimate and CVs of the rate for people 65 years of age and older illustrates the concern. In 2010, the estimated incidence rate was 0.9. In 2011, the incidence rate almost doubled to 1.7. However the CVs in 2010 were 33 percent, giving a confidence interval of [0.249, 1.55]. Thus a data user could not tell whether there was a major increase in rape and sexual assault for older people, or whether the estimated change was due to random variation. By marital status, people who were married but separated experienced the highest estimated victimization rates, along with year-to-year instability in the estimates. The CVs for these rates in 2010 and 2011 were around 20 percent. Again, data users have said that the year-to-year variability is a particular issue for their use of the estimates.

Table 7-3 shows similar data by geographic areas. For serious violent

TABLE 7-2 NCVS Victimization Rates and Coefficients of Variation (CVs) for Serious Violent Crimes, per 1,000 People (12+ years) for Selected Subpopulations

	Estimated Serious Violent Crime			CVs (%)		
Category	2002	2010	2011	2002	2010	2011
Total	10.0	6.6	7.2	5.0	6.1	5.6
Gender						
Male	10.4	6.4	7.7	6.7	9.4	9.1
Female	9.5	6.8	6.7	7.4	8.8	9.0
Race/Ethnicity						
White	8.6	5.8	6.5	5.8	8.6	7.7
Black	17.8	10.4	10.8	9.6	12.5	13.0
Hispanic	12.3	6.7	7.2	11.4	13.4	13.9
American Indian/Alaskan Native	14.3	47.3	12.6	47.6	24.1	50.8
Asian	3.4	2.3	2.5	32.4	35.8	36.0
2 or more		17.7	26.2		29.4	24.0
Age						
12-17	17.0	11.7	8.8	10.0	12.8	15.9
18-24	24.7	17.0	16.3	8.1	10.0	11.0
25-34	12.3	7.1	9.5	9.8	12.7	11.6
35-49	7.6	5.6	7.0	9.2	12.5	11.4
50-64	4.4	3.7	4.3	13.6	16.2	14.0
65 and older	1.8	0.9	1.7	22.2	33.3	23.5
Marital Status						
Never married	16.1	11.9	11.7	6.8	7.6	8.5
Married	5.7	2.2	3.7	8.8	13.6	10.8
Widowed	4.4	3.0	0.7	25.0	30.0	57.1
Divorced	10.9	11.2	9.2	13.8	13.4	15.2
Separated	34.8	18.8	26.4	15.2	20.7	18.9

SOURCE: Data from *Criminal Victimization, 2011* (Bureau of Justice Statistic, 2012a).

TABLE 7-3 NCVS Victimization Rates and Coefficients of Variation (CVs) for Serious Violent Crimes, per 1,000 People (12+ years) by Geographic Areas

	Serious Violent Crime			CVs (%)		
Area	2002	2010	2011	2002	2010	2011
Total	10.0	6.6	7.2	5.0	6.1	5.6
Region						
Northeast	7.1	6.8	6.4	11.3	13.2	14.0
Midwest	11.5	7.6	7.8	8.7	10.5	11.5
South	10.8	5.4	6.5	7.4	11.1	10.8
West	9.5	7.5	8.4	9.5	10.7	9.5
Location of residence						
Urban	15.2	9.5	9.7	7.2	8.4	9.3
Suburban	7.8	5.5	5.7	7.7	9.1	8.8
Rural	7.9	4.7	6.7	11.4	14.9	14.9

SOURCE: Data from *Criminal Victimization, 2011* (Bureau of Justice Statistic, 2012a).

crimes, the CVs for regional estimates are generally 10 percent or somewhat higher. The estimated rates for urban and suburban areas have CVs of less than 10 percent, but the CVs for households in rural areas are about 15 percent, and there is considerable fluctuation in this estimate from year to year.

Data users have a great deal of difficulty establishing temporal trends with wide fluctuations due to sampling error in the annual estimates. The executive director of the Rape, Abuse and Incest National Network (RAINN), the nation's largest anti-sexual violence organization, has indicated that the problem of inadequate sample size on the NCVS is very serious from the data user perspective (Berkowitz, 2011). He indicated that wildly fluctuating year-to-year numbers is one of his major concerns with the NCVS because it makes difficult the analysis of trends across years.

Pooling NCVS data across years is currently the only way to look at subpopulation estimates of rape and sexual assault. Many data users use this technique to examine subgroup risk for rape and sexual assault. For example, Lauritsen (2012) combined up to 15 years of data to produce estimates and confidence intervals. Berkowitz (2011) also reported combining several years of data to follow rate changes over time. This technique can prove very useful, but it has potential pitfalls, one of which is that it can obscure important year-to-year changes. The finer (more disaggregated) the subgroups that are included in an analysis, the greater the number of years that must be combined to obtain a sufficient number of positive responses to stabilize the estimates.

CONCLUSION 7-1 The National Crime Victimization Survey, which is designed as an omnibus victimization survey, is efficient in measuring the many types of criminal victimizations across the United States, but it does not measure the low incidence events of rape and sexual assault with the precision needed for policy and research purposes. Comparisons across subgroups and years are particularly problematic.

The Effect of Series Victimization on Estimates

Series victimization is defined on the NCVS as the situation when a single respondent reports six or more separate but similar victimizations over the reference period but is unable to recall these events individually or describe them separately in detail to the interviewer (see Chapter 4).[2] Table 7-4 shows that approximately 6 percent of the reports of rape and sexual assault are identified as series victimizations, a larger percentage than for other crimes. Lauritsen et al. (2012) provide analysis of the distribution of series victimizations for all violent crimes, but they are not often able to isolate results for rape and sexual assault. The authors found evidence of response error in these reports because these victims (Lauritsen et al., 2012, p. 13)

> [had] difficulty recalling exactly how many times violent victimizations occurred within a 6-month reference period. The observed patterns of response clustering indicated that many victims provided estimates of the number of times the victimizations occurred rather than counting directly from memory.

Thus, when an individual is victimized so many times during a 6-month period that he or she has difficulty recalling individual incidents, that respondent may also have difficulty providing an accurate count of the number of incidents that happened and whether the incidents occurred within the reference period. Lynch, Berbaum, and Planty (2002, p. 23) further speculated about another potential measurement error problem that may exist in this category:

> [S]eries incidents in a large part may be an artifact of Census Bureau procedures. More specifically, multiple events may be treated as a series event when the respondent can clearly recall and report on these incidents, simply because it is easier for the interviewer to complete a single incident form, as opposed to multiple incident forms.

[2]If a respondent suffers multiple victimizations and is able to recall the events individually, then the situation is not classified as a series victimization, and each victimization is recorded separately.

TABLE 7-4 Victimizations Reported as Series Victimizations in the NCVS, by Type of Crime, 1993-1999 and 2000-2009, as Percentage of All Victimizations Reported

Category of Victimization	1993-1999	2000-2009
Rape and sexual assault	6.3	5.7
Robbery	2.9	2.5
Aggravated assault	4.6	3.1
Simple assault	6.9	4.3
Personal larceny	0.3[a]	0.7[a]
Burglary	1.4	1.0
Motor vehicle theft	0.3	0.2[a]
Theft	1.1	0.7

NOTE: Table shows the number of incident reports recorded under "series victimization" procedures (which will include multiple victimizations) as a percentage of all incident reports recorded.

[a]Interpret with caution; estimate based on 10 or fewer sample cases, or coefficient of variation is greater than 50 percent.

SOURCE: Lauritsen et al. (2012, p. 3, Table 2).

Lauritsen et al. (2012) found some supportive evidence for this conjecture.[3]

From a statistical point of view, series victimization procedures create outlier problems for estimation. In general, outlier problems can be caused by large estimation weights, large outlying data values, or moderate values. Estimation weights for the NCVS are fairly large. When estimating rape and sexual assault (a low-incidence item in the NCVS data), the data values are generally zero (no rape or sexual assault reported). When rape or sexual assault is reported as a series, the data value can be quite high.[4] Under the new procedures the value is truncated at "10" for individuals reporting more than 10 incidents in a single series. Even with the truncation, these outliers (representing only 6 percent of the positive responses to rape and sexual assault) tied to the NCVS weights have a substantial impact on the estimates and the standard errors of those estimates, with both increasing fairly substantially. Fortunately, the statistical literature is fairly well developed in the areas of detecting and adjusting for outliers, and some of the developed techniques (adjusting the weights, the data value, or both)

[3]By the definition of series victimization, the respondent must report six or more similar victimizations during the reference period for which she or he cannot recall the separate details. However, Lauritsen et al. (2012) found examples of records categorized as a series victimization in which the respondent identified fewer than six victimizations.

[4]Lauritsen et al. (2012, p. 10) reported a maximum value of "750" incidents in series has been reported for serious violent crimes. They did not report the maximum value reported for rape and sexual assault.

may be appropriate for use in measuring rape and sexual assault (see, e.g., Barnett and Lewis, 1994; Hodge and Austin, 2004; Pedlow et al., 2010).

Until 2011, NCVS deleted these outliers for the purpose of estimates reported in *Criminal Victimization* (although they counted a series as a single victimization, rather than deleting, in some special reports). The effect was to heavily suppress the larger numbers that were reported by ignoring these multiple victimizations. This process added to a potential underestimation of victimizations (Planty and Strom, 2007).

Beginning in 2011, BJS stopped deleting these outliers. Instead, reported series victimizations are now directly included in the estimates with no additional adjustment unless more than 10 victimizations are reported in one series. Reported values greater than 10 are truncated to the value of 10. BJS has made the change retroactively back to 1993 in its online NCVS database.[5]

The effect of changing the method for handling these outliers in the estimates of rape and sexual assault is huge (see Figure 7-1 and Table 7-5). Across the past 18 years, this change in methodology increased the estimates of incidents of rape and sexual assault by an average of 52 percent per year, and it increased the estimates of incidence rate by 55 percent. The estimates (number of victimizations) also fluctuated more from year to year. The change ranged from a low of zero percentage change in 2007 (there were no series victimizations reported) to a high of 143 percentage change in 2009.

An important question is how large are these changes relative to the overall sampling error of the estimates. Figure 7-1 shows confidence intervals and thus allows a comparison as to whether the 95 percent confidence intervals for the two estimates ("with series victimization" and "without series victimization") generally overlap. Of the 18 years depicted in the graphics, the confidence intervals for the two estimates overlapped seven times and did not overlap eight times. There were 3 years (1996, 2003, and 2008) in which the end points of the two bands touched with minimal overlap.

Figure 7-2 displays the percentage of each estimate that is directly attributable to series victimization outliers. It shows that the effect is large and very volatile across years. In 2009, nearly 60 percent of the estimate of the number of rape and sexual assault victimizations was due to series victimizations. Combining the two graphics (Figure 7-1 and Figure 7-2) together, we see that the confidence intervals overlap in years when the series victimizations make up less than 30 percent of the total estimate.

The panel did not have the time and resources to examine specific alternative outlier adjustment procedures. Thus we address the issue in

[5]The panel found out about this change late in its work and so was unable to analyze its significance as thoroughly as we would have liked.

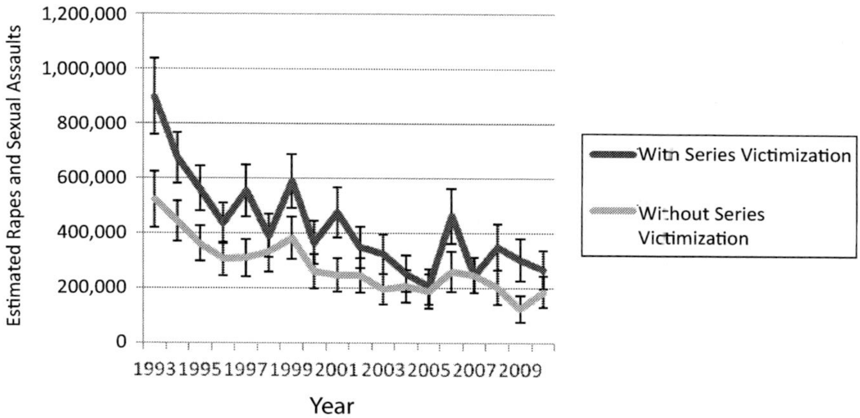

FIGURE 7-1 Estimates and confidence intervals for rape and sexual assault, with and without series victimizations, NCVS, 1993-2010.
SOURCE: Bureau of Justice Statistics (1994, 1995, 1996, 1997, 1998, 1999, 2000, 2001, 2002a, 2003, 2004, 2005, 2006, 2007, 2008a, 2009, 2010, 2011, n.d.-a).

more general terms. It is clear that a subpopulation is at risk for being repeatedly raped and sexually assaulted over a relatively short period of time. This is a statistically rare subpopulation within the population of all victims of rape and sexual assault, which already has a low incidence rate. The panel believes that this subpopulation needs to be better understood, and its victimizations should be accounted for in BJS statistics. However, the panel believes that a more sophisticated approach than currently used may be needed to properly represent series victimizations in the estimates of incidence rates for rape and sexual assault.

CONCLUSION 7-2 Records identified as series victimizations create an outlier problem in the estimation process for the National Crime Victimization Survey. The current method for handling series victimization, although an improvement over the method used until 2011, allows these relatively rare reports to have a large impact on the national estimates of rape and sexual assault and creates large year-to-year volatility.

TABLE 7-5 Effect of Series Victimizations on National Crime Victimization Survey Estimates of Rape and Sexual Assault, by Year

	Number of Victimizations				
Year	Counting Series Victimization	Standard Error Series	Ignoring Series Victimization	Standard Error Ignoring Series	Percentage Change in Estimate
1993	898,239	71348	521,223[a]	51573	72
1994	674,291	47198	433,509[a]	36858	52
1995	563,249	42418	363,527[a]	32830	55
1996	437,198	37345	307,100	30430	42
1997	553,523	48035	311,110	34409	78
1998	391,101	39935	332,500	36103	18
1999	591,460	49907	383,170	38340	54
2000	366,747	39485	260,950	32152	41
2001	476,578	46216	248,250	31043	92
2002	349,805	38253	247,730	31352	41
2003	325,311	36759	198,850	27884	64
2004	255,769	33339	209,880	29887	22
2005	207,760	32551	190,592[b]	31032	9
2006	463,598	50305	260,940	36990	78
2007	248,277	32924	248,277	32924	0
2008	349,691	42837	203,830	31719	72
2009	305,574	39443	125,910	24079	143
2010	268,574	36057	188,380	29399	43
2011	248,803	34800	217,331	32616	14
Average Percentage Change Across Years					52

[a]The estimates published in *Criminal Victimization, 1993, 1994, 1995* were revised in 1996 to reflect a methodology change to estimate victimizations for the "collection year" rather than the year in which the victimization occurred.

[b]Based on errata issued June 16, 2011.

Incidence Rate per 1,000 People (12+ years)		
Counting Series Victimization	Ignoring Series Victimization	Percentage Change in Estimate
4.3	2.5	87
3.2	2.1	60
2.6	1.7	63
2.0	1.4	43
2.5	1.4	79
1.8	1.5	20
2.6	1.7	53
1.6	1.2	33
2.1	1.1	91
1.5	1.1	36
1.4	0.8	75
1.1	0.9	22
0.8	0.8	0
1.9	1.1	73
1.0	1.0	0
1.4	0.8	75
1.2	0.5	140
1.0	0.7	43
		55

SOURCES: Bureau of Justice Statistics (1994, 1995, 1996, 1997, 1998, 1999, 2000, 2001, 2002a, 2003, 2004, 2005, 2006, 2007, 2008a, 2009, 2010, 2011, n.d.-a).

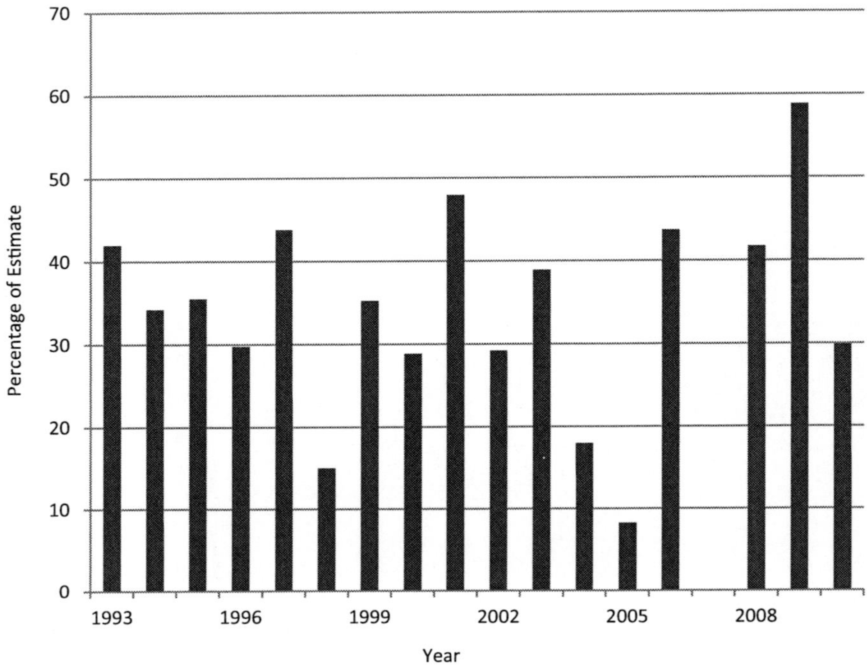

FIGURE 7-2 Effect of including series victimization in rape and sexual assault estimates in the National Crime Victimization Survey.
SOURCE: Panel-designed graphic using data from Bureau of Justice Statistics (1994, 1995, 1996, 1997, 1998, 1999, 2000, 2001, 2002a, 2003, 2004, 2005, 2006, 2007, 2008a, 2009, 2010, 2011, n.d.-a).

FRAME ERROR

A sampling frame is the source material (such as a listing of people, addresses, or counties) from which a sample is selected. Errors can arise in the process of constructing, maintaining, or sampling from a frame. A quality sampling frame should provide complete (or nearly complete) coverage of the target population. The operational task of building a sampling frame is difficult and susceptible to error; consequently, steps in that process need to be operationally workable and monitored for quality. The frame should provide a base for reasonable response rates and accurate collection of data. If used for ongoing surveys, the frame must be maintained to meet specified standards.

The NCVS uses several sampling frames that are built and maintained by the Census Bureau and which the Bureau uses for a number of large national surveys. In particular, the NCVS uses a two-stage design similar to that used in the Census Bureau's Current Population Survey. The primary frame is the area frame built for the decennial census. PSUs are selected from this frame, consisting of small groups of neighboring counties that include individual counties, groups of counties, or large metropolitan areas. In the second stage of sampling, the NCVS uses four secondary frames within selected PSUs.

A secondary frame comes from the Census Bureau's Master Address File (MAF), which is developed through address listing activities in support of the decennial census. For the NCVS, it is used within selected PSUs. Information from a decennial census is generally not available for use in the NCVS sample until 5 years after a census year (e.g., 2005), and it then is used for the next 10 years. Because the MAF is not comprehensively updated during this time and can become out-of-date, the Census Bureau uses a frame based on new building permits to supplement the existing MAF between censuses. Another secondary frame, the "area" listings, is generated from periodic canvassing of selected census blocks within the PSUs: they consist of identified residential units that are not on the other secondary frames. This process of supplemental listings is helpful but still leaves a basic problem: it slows but does not eliminate the inevitable coverage lapses and inefficiencies of the MAF over time.

An additional secondary frame developed for the census process and used by the NCVS is a "noninstitutionalized group quarters" listing. This frame includes such facilities as college residence halls, halfway houses for substance abuse, homes for the developmentally disabled and for the physically handicapped, religious group quarters, agricultural workers dormitories, vocational training residence facilities, and more. It does not include institutions such as prisons. A study of residence rules in the decennial censuses (National Research Council, 2006, p. 6) found issues with the construction of the group quarters frame and its enumeration:

> [A]s implemented in the 2000 and recent censuses, group quarters enumeration is unacceptably bad. Failure to reconcile the group quarters roster with the MAF contributed to a host of census errors. Group quarters frames were constructed without sufficient standardization and awareness of diversity in housing unit and group quarters stock . . . the challenge of collecting even the basic census items from group quarters' populations remains.

Another report from the National Research Council (2012) looked at improving the measurement of people living in group quarters on the American Community Survey (ACS), a survey that is sampled from this same frame.[6] Its title—*Small Populations, Large Effects*—could also be used to describe the measurement of sexual violence to people who live in group quarters. Group quarters residents are likely to have higher, or at least different, risks of such violence. Thus, they are very important to the population estimates, but the number of group quarters selected in the sample is very small. The National Research Council's report (2012, p. 5) on the ACS concluded:

> [The estimates] must rely on a sample of what is a small and very diverse population, combined with limited funding available for survey operations, makes the ACS GQ [group quarters] sampling, data collection, weighting and estimation procedures more complex and the estimates more susceptible to problems stemming from these limitations.

The report provided several recommendations, including that the "Census Bureau should give high priority to developing a detailed and systematic operational plan . . . for a group quarters address updating system" (National Research Council, 2012, p. 5).

CONCLUSION 7-3 Because the decennial Master Address File, which is used for selecting the main second-stage sample of the National Crime Victimization Survey, is about 15 years old before it is completely updated, there is potential for incompleteness and inefficiency in the frame. The use of a supplemental frame of new building permits and a periodic "area" canvassing of some primary sampling units are helpful, but not a complete solution.

CONCLUSION 7-4 The frame for the ancillary listing of group quarters, which is an important part of the secondary sample for the National Crime Victimization Survey because their residents may be at higher risk for sexual violence, is seriously flawed in terms of both the building and enumeration of this secondary frame.

PROCESSING ERROR

The panel reviewed the processing code that categorizes reported incidents and developed a flow chart of that classification coding. BJS has created eight different categories that encompass the sexual-related incidents that they define as criminal:

[6]The American Community Survey, which is continuous, has replaced the former "long form" of the decennial census.

1. completed rape,
2. attempted rape,
3. sexual assault with serious assault,
4. sexual assault with minor assault,
5. sexual assault without injury,
6. unwanted sexual contact without force,
7. verbal threat of rape, and
8. verbal threat of sexual assault.

Incidents reported on the NCVS that are classified in one of these categories are considered "criminal victimizations." The process of classification is complex. Figure 7-3 shows a flow chart for this classification for the four most serious victimizations listed above. It is unclear how often classification errors occur, or how many victimizations fall initially into an "unclassified code" because they cannot be classified through an automated process.

The current screening questionnaire used on the NCVS uses a series of cue screening questions that talk about various life domains such as work and leisure, and trigger memories of victimizations that may have occurred in these domains. A respondent may report a victimization at any point during the screening questionnaire or when answering questions in an incident report. The panel has seen no data that shows the relative rate of reporting of sexual victimization in various parts of the screener questionnaire (or within an incident report for another type of victimization). This would be very interesting data to see, and deserves greater transparency.

The NCVS, like any major survey, includes an extensive edit process. Cleaning up data that have been collected is a major part of the total data collection process. The edit process used by BJS for the NCVS is not included in the description of NCVS methodology. Thus, it was not possible for the panel to assess it, and it is similarly not possible for data users to understand this important assessment of the NCVS.

CONCLUSION 7-5 The Bureau of Justice Statistics does not provide public information on the edit process in the National Crime Victimization Survey, although processing and editing errors are an important part of any major survey data collection. The lack of transparency about these processes makes it difficult for data users to fully understand the survey's estimates.

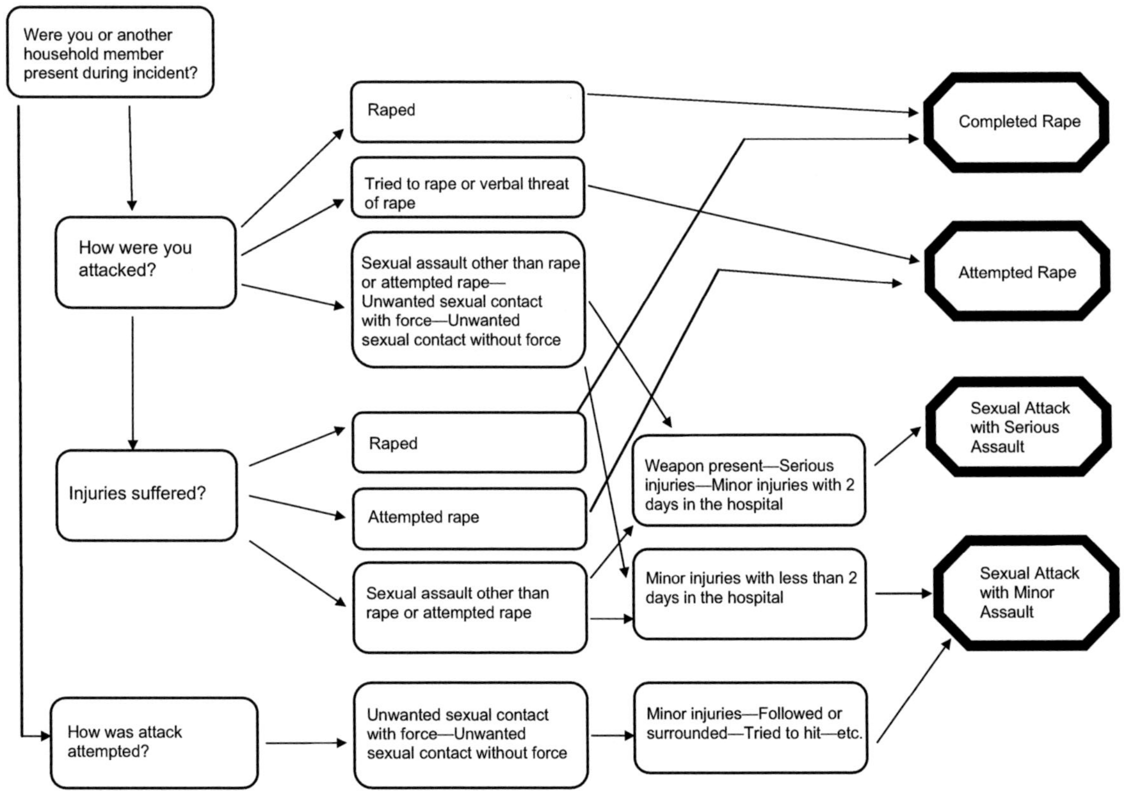

FIGURE 7-3 A flow chart for crime classification of reported victimizations on the National Crime Victimization Survey for the four most serious victimizations: completed rape, attempted rape, sexual attack with serious assault, and sexual attack with minor assault.

8

Potential Sources of Error: Nonresponse, Specification, and Measurement

This chapter continues the analysis in Chapter 7 of potential sources of error in the National Crime Victimization Survey (NCVS), covering nonresponse error, specification error, and measurement error.

NONRESPONSE ERROR

Nonresponse error in surveys arises from the inability to obtain a useful response to all survey items from the entire sample. A critical concern is when that nonresponse leads to biased estimates. Nonresponse bias is a product of the difference between respondents and nonrespondents on a particular measure and the size of the nonresponse population. A lower response rate increases the potential for greater nonresponse bias, but when the data are missing at random, a lower response rate will neither create nor increase nonresponse error.

The NCVS, like most federal household surveys, is voluntary and not required by law. The challenges facing today's federal household surveys were recently summarized by the National Research Council (2013a, p. 68):

> [They] include maintaining adequate response from increasingly busy and reluctant respondents. More and more households are non-English speaking, and a growing number of higher income households have controlled-access residences. . . .Today's household surveys face confidentiality and privacy concerns, a public growing more suspicious of its government, and competition from an increasing number of private as well as government surveys vying for the public's attention.

These challenges mean that maintaining a high level of response on a large voluntary national survey is difficult. This section examines the nonresponse profile of the NCVS, looking at both the level of nonresponse and its potential effect on the measured rate of sexual victimization.

Nonresponse can arise at several points in the process of sample recruitment. In the NCVS, the household address is selected, and then each household member (12 years of age and older) is asked to complete the survey. Nonresponse on a questionnaire (unit nonresponse) can occur at two stages. Household nonresponse occurs when no one living at the selected housing unit responds in the data collection wave. Person-level nonresponse occurs when some eligible persons in the household respond and some do not respond. In addition, a household or person may respond on some waves but not on all waves. In the NCVS, a household responding to at least one wave of the NCVS is counted as a household respondent for the survey. Likewise, a person who is interviewed in one or more waves is called a person respondent. Finally, item nonresponse (as opposed to unit nonresponse) can also occur for person respondents when some questions on the questionnaire were not completed that should have been completed.

This section looks in more depth at the person-level nonresponse at both the unit and item level.

Unit-Level Nonresponse

The NCVS has maintained a moderately high level of survey (unit) response at both the household level and the person level (see Table 8-1). In 2011, 79,800 households participated in the NCVS, representing a 90 percent household response rate for the year.[1] The person-level response rate (most important for victimization rates) was 88 percent in 2011. Response rates have decreased several percentage points over the decade, but not substantially (see Table 4-2 in Chapter 4). These response rates are consistent with several other important federal household surveys in 2011.[2]

Nonresponse in a survey may be "missing at random" (MAR), mean-

[1]It appears that the Census Bureau is defining the housing unit response rate as the number of housing units that participated at one or more waves during the year divided by the number that should have participated during the year. This is an inflated number because if a housing unit participated in January but not in July (or vice versa), then it is still counted as a respondent for the year. A better measure of the response rate is the number of times housing units participated divided by the number of times housing units were eligible to participate. We believe this response rate calculation is a better indicator of the potential for (or risk of) nonresponse bias than the current way the response rate is calculated.

[2]For comparison, in 2010 the Current Population Survey had monthly household response rates of 91-93 percent; the American Community Survey had a household response rate of 98 percent; the National Health Interview Survey had a household response rate of 82 percent; and the Consumer Expenditure Survey had a household response rate of 73 percent.

TABLE 8-1 National Crime Victimization Survey Response Rates for Households and Individuals

	Household Level		Person Level		
Year	Responding	Response Rate	Total Persons	Responding Persons	Response Rate
1996	45,000	93	NA	85,330	91
1997	42,910	95	NA	79,470	90
1998	43,000	94	NA	78,900	89
1999	43,000	93	204,915	77,750	89
2000	43,000	93	207,800	79,710	90
2001	44,000	93	208,598	79,950	89
2002	42,000	92	203,061	76,050	87
2003	42,000	92	201,388	74,520	86
2004	42,000	91	202,771	74,500	86
2005	38,600	91	181,009	67,000	84
2006	38,000	91	179,717	67,650	86
2007	41,000	90	170,869	73,650	86
2008	42,093	90	155,704	77,852	86
2009	38,728	92	157,796	68,665	87
2010	40,974	92	NA	73,283	88
2011	79,800	90	NA	143,120	88

SOURCES: Data from Bureau of Justice Statistics (1997, 1998, 1999, 2000, 2001, 2002a, 2003, 2004, 2005, 2006, 2007, 2008a, 2009, 2010, 2011, 2012).

ing that the decision not to respond on the survey is unrelated to key study outcome measures, such as crime victimization, and that reweighting of the responding units may suffice to adjust for the missing data. The presence of this type of nonresponse, when appropriately reweighted, does not cause a bias, but does reduce sample size and increase sampling error. Other unit nonresponse is judged to be "not missing at random" (NMAR) and thus is more of a problem because it can produce bias in the estimates as well as increase the sampling error. If the nonresponse varies with key outcome measures and their covariates (such as race, income, or geographic area), then the nonresponse may be MAR within groups formed based on these covariates. In this case, reweighting might be done within the groups, thus reducing potential nonresponse bias.

Because of the panel nature of the NCVS, considerable information is known about the demographics of selected households and individual household members if they respond at least once over the 3-year life of the panel. The Bureau of Justice Statistics (BJS) uses this information for much more than just adjusting for nonresponse. For example, one adjust-

TABLE 8-2 Survey-Level Nonresponse on the National Crime Victimization Survey Judged to Be Missing at Random (MAR), by Subgroups

Subgroup	Percentage of Nonresponses Judged MAR	Total Counts of Survey-Level Nonrespondents Judged MAR
All	81.10	2,762
Male	84.04	1,327
Female	83.43	1,435
Black	84.81	469
Other	80.43	2,294
25 years of age and younger	84.11	323
25 years of age and older	83.74	2,441

SOURCE: NORC at the University of Chicago (2009, p. 19, Table 2.5).

ment is to "inflate sample point estimates to known population totals to compensate for survey nonresponse and other aspects of the sample design" (Bureau of Justice Statistics, 2008b, p. 12). (See Chapter 4 for more detail on this and other adjustments.)

The success of the BJS adjustment processes in addressing potential unit-level nonresponse bias in the NCVS was examined by NORC at the University of Chicago (2009) in an extensive study with several parts. In one part, NORC conducted a capture-recapture analysis across panel waves to obtain relative counts of different categories of nonrespondents. This technique separates the chronic nonresponders across the 3 years from the occasional and frequent responders, hypothesizing that the chronic nonresponders were potentially NMAR. Based on the above assumption as to which respondents were NMAR, the NORC report estimates that 81 percent of the nonrespondents are not chronic nonresponders and may be assumed to be MAR (see Table 8-2). Using the term "ignorable" for MAR and nonignorable for NMAR nonresponse, the report (NORC at the University of Chicago, 2009, p. 16) concludes:

> Overall, more that 80 percent of the nonresponses in NCVS can be regarded as "ignorable." Proportionately, more nonresponses by male, black, and young (age 25 or less) eligible interviewees are ignorable. The largest of variation occur for the race/ethnicity, with eligible black interviewees having proportionately more ignorable nonresponses (84.81% vs. 80.43%).

NORC points out that its techniques did not allow analysis of nonresponse in the first round of the panel.

In a subsequent part of the report, NORC developed log linear models to predict response disposition for key subgroups. The models examined "easy versus hard" responder characteristics. Finally, NORC made county-level comparisons between the statistics from the Uniform Crime Reports and the NCVS pooled across year. The report's conclusion (NORC at the University of Chicago, 2009, p. 47) is "little evidence for nonresponse bias after the first round of the survey. . . . The within unit nonresponse is weight adjusted to age and race controls in the NCVS and these seem to be the categories that are the main drivers in any potential nonresponse bias."

The panel has important reservations about some of the NORC analysis and conclusions. The capture-recapture analysis is based on the assumption that individuals who respond at least once but not routinely on the NCVS are MAR. This assumption appears to go untested and yet underpins NORC's overall analysis. Another limitation is that the logistic modeling techniques used in the study only looked at a few standard demographic characteristics. Finally, it is unclear whether this broad look at nonresponse on the NCVS paints the same picture as would an analysis of the subpopulations that are at greater risk for sexual violence.

CONCLUSION 8-1 The overall unit response rates, as calculated, on the National Crime Victimization Survey are moderately high and have been reasonably stable over the past 10 years. Although an independent analysis concluded that the methods that the Bureau of Justice Statistics uses to adjust for nonresponse appear to provide a satisfactory correction for nonresponse bias at the unit level, our panel has reservations about that analysis and remains concerned that there may be a nonresponse bias related to sexual victimization.

Panel Attrition

Panel attrition is a response pattern in surveys with multiple waves of data collection in which a respondent's propensity to respond decreases over these waves. Because the NCVS is a panel survey with seven waves of data collection over 3 years, it is important to examine the nonresponse pattern across waves. There are many reasons that an individual may attrite, including deciding to quit reporting, not being available during the data collection period, or moving to a different address.

BJS does not provide NCVS response rates by wave. To get some sense of attrition rates, the panel calculated unweighted response rates (at the person level)[3] using data for 2007-2008 by time in sample (see Figure 8-1).

[3]The first wave person-level response rate is the proportion of persons participating at first wave among sampled, eligible persons at first wave. The person-level attrition rate at wave $t > 1$ is the proportion of persons who participated at first wave who also participated at wave t.

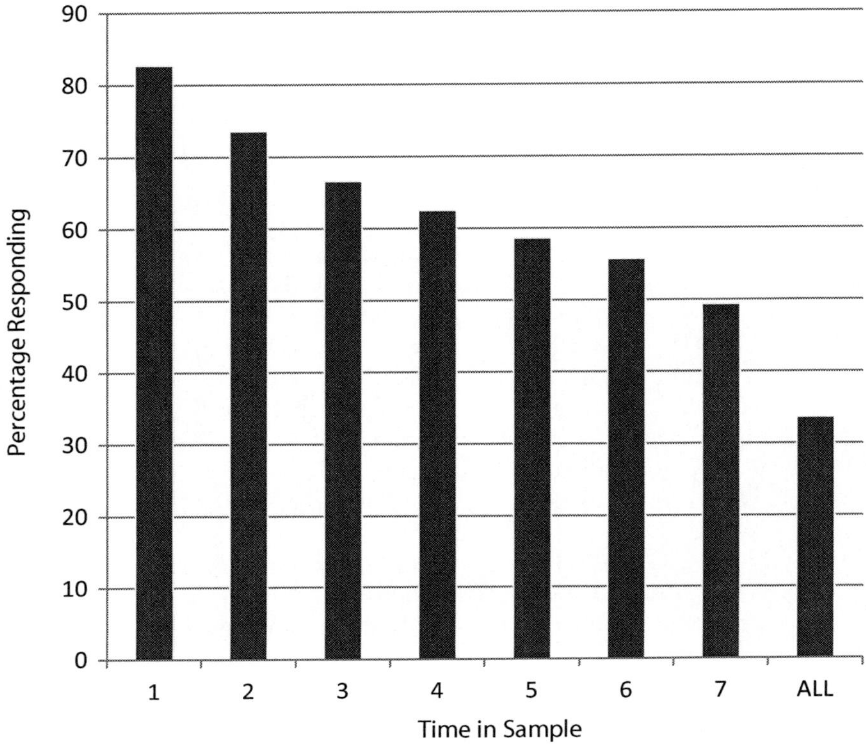

FIGURE 8-1 National Crime Victimization Survey person-level attrition rates (unweighted) for the period 2007-2008.
NOTES: Time in sample (TIS) 1 is the response rate at the initial wave, TIS 2-7 is the response rate given response at TIS 1, and ALL is the proportion of eligible person responding to all seven waves.
SOURCE: Data from National Crime Victimization Survey, 2007-2008.

These attrition rates were calculated at the person level using linked longitudinal files. One can see substantial attrition in response rates over time, with less than half the sample responding in all waves.

The NORC at the University of Chicago (2009) report provides insight into this panel attrition by subgroups. The report's analysis is based on the total number of waves in which a respondent participated, without an ordering of those waves over time. Looking at the age of the respondents, the analysis found that younger respondents participated in fewer waves than did older respondents (see Figure 8-2).[4] Approximately 15 percent of

[4]Data included only individuals who had participated in the first wave.

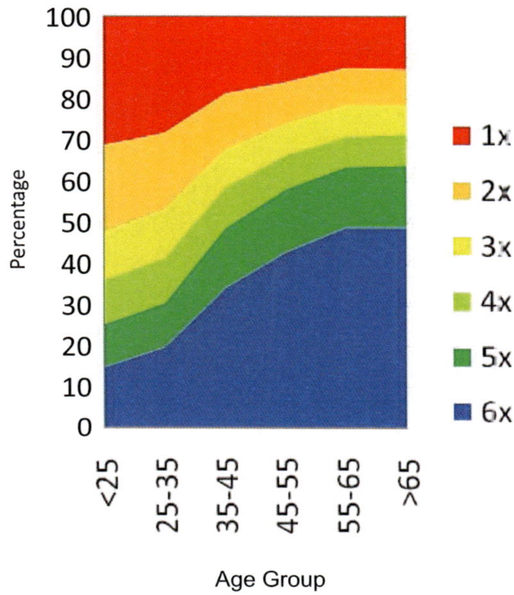

FIGURE 8-2 Participation in National Crime Victimization Survey waves by respondents' ages, 2005-2006.
SOURCE: NORC at the University of Chicago (2009, p. 29, Chart 3.3).

respondents 25 years of age and younger participated in all seven waves; in contrast, approximately 45 percent of respondents 55 years of age and older did so. And as can be seen in the figure, nearly 30 percent of respondents 25 years of age and younger did not participate after the first wave.

The NORC at the University of Chicago (2009) report also looks at response by household structure (see Figure 8-3). Individuals living as couples (couple only, couple with kids and family, couple with others) responded in more waves than did individuals who were not identified as being part of a couple (male with relatives, male with others, female with relatives, female with others).

The results shown in both of these figures provide particular concern for the estimation of rape and sexual assault because the low responders—particularly young people and females who are not part of a couple—appear to be more at risk for being victims of those crimes. In a multivariate analysis of subgroup risk among females for rape and sexual assault (Lauritsen, 2012), younger people (in the age groups 12 to 17, 18 to 34, and 35 to 49) have a higher odds ratio than do older (50+) individuals (see Table 8-3). Females who are not part of a couple (widowed, divorced, separated, and

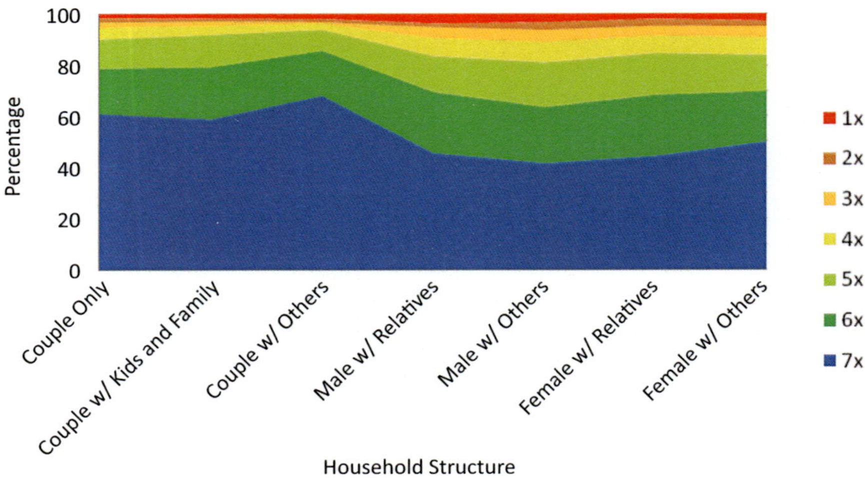

FIGURE 8-3 Participation in National Crime Victimization Survey waves by family structure, 2002-2006.
SOURCE: NORC at the University of Chicago (2009, p. 28, Chart 3.2).

never married) have a higher odds ratio than do married women. Planty et al. (2013) provide similar results.

Thus, attrition rates are higher in several subgroups that appear to be at higher risk for sexual violence. It is unclear whether this is a related effect. One could argue that someone who has been sexually victimized may be

TABLE 8-3 Risk for Rape and Sexual Assault for Females, by Age and Marital Status, National Crime Victimization Survey, 1994-2009

	Odds Ratio	95% Confidence Interval	Significance
Age (in comparison with 50+)			
35 to 49	4.6	[3.31, 6.38]	*
18 to 34	8.7	[6.21, 12.22]	*
12 to 17	9.23	[6.31, 13.40]	*
Marital status (in comparison with married)			
Widowed	2.48	[1.43, 4.29]	
Divorced	5.56	[4.44, 6.96]	*
Separated	10.51	[7.89, 14.00]	*
Never married	3.90	[3.12, 4.87]	*

*The odds ratios for rape and sexual assault are significantly greater than the odds ratios for other forms of serious violence.
SOURCE: Lauritsen (2012, Table 6).

less willing to respond on the next NCVS, knowing that questions regarding victimization will be asked. Similarly, one could argue that someone who has been sexually victimized may be more likely to move to a safer neighborhood, and thus no longer be an eligible respondent. The panel did not find data that could answer this question definitively, but there appears to be potential for a nonresponse bias that could contribute to underreporting of these victimizations.

> **CONCLUSION 8-2 There appears to be notable panel attrition over the 3 years in the National Crime Victimization Survey (NCVS). This attrition is particularly problematic for estimating rape and sexual assault because some people at greater risk for being victimized by these crimes—young people and females not living as part of a couple—are also some of those most likely to drop out before the seven waves of the NCVS have been completed.**

> **CONCLUSION 8-3 Although the Bureau of Justice Statistics publishes annual response rates for the National Crime Victimization Survey (NCVS), the published data do not include important details of response, such as mode of data collection and attrition rate. Such details are needed by data users for a thorough assessment of the quality of NCVS estimates.**

Item Nonresponse

Item nonresponse occurs when a respondent completes a substantial portion of a questionnaire (enough to count the interview as "complete") but does not provide answers to certain key items. The panel could not find an analysis of item nonresponse on the NCVS in general, nor one specifically for the questionnaire items regarding rape and sexual assault. Without such analysis, the panel relied on its collective experience and judgment about item response for key questions regarding sexual victimization.

There is considerable evidence in survey research that respondents are reluctant to answer socially undesirable questions (Bradburn, 1983; Schaeffer, 2000; Tourangeau and Smith, 1996; Tourangeau and Yan, 2007). (See also the section on "Questionnaire" in this chapter.) The panel thinks that item "refusals" on these particular socially undesirable questions would be difficult to identity. If a respondent does not want to report a rape or sexual assault, or to talk about such an assault, then he or she is more likely to answer NO to the appropriate screening questions (he or she was not victimized) rather than more directly refusing to answer the question. In fact the screening questionnaire (Bureau of Justice Statistics, n.d.-d) has only check boxes for YES or NO for these questions, and no response box for "refused" or "don't know." Thus, these item refusals are most likely disguised as legitimate zeros (there was no victimization).

Panel surveys may create an additional nuance regarding item nonresponse. After going through one or more waves of the survey, a respondent learns that answering YES to a screening question will lead to a range of additional questions regarding the specific incident. Surveys with this repeated pattern, and especially those with the pattern repeated across multiple waves, are subject to "satisficing": a respondent provides an answer (perhaps NO to a screening question) that moves the interviewer on to the next question, without necessarily being an accurate or complete response. This respondent conduct is hard to detect and measure, but the panel thinks it is likely that satisficing is occurring on the NCVS.

CONCLUSION 8-4 The panel believes it is likely that item refusals on questions about sexual victimization on the National Crime Victimization Survey may be recorded as if they were "no" response rather than item nonresponse when a respondent does not want to report a victimization. Another possibility is for a respondent to sometimes answer "no" on screening questions simply to avoid additional questions in the survey.

SPECIFICATION ERROR

For any survey, its intended purpose and concepts must be clearly defined in order for survey instruments and procedures to accurately translate those concepts into the collection of data. In surveys, specification error may occur when there is a mismatch between what the survey is measuring and what it is intended to measure.[5] As defined by Biemer (2010, p. 31): "specification error pertains specifically to the problem of measuring the wrong concept in a survey, rather than measuring the right concept poorly." This section examines a key concept associated with the NCVS to see if it is clearly defined and consistent between the survey's purposes and processes.

This key concept is to identify if and when a respondent has been the victim of a rape or sexual assault. BJS has developed a clear definition of what the survey is intended to measure (see Box 8-1). In the omnibus screener that is currently used in the NVCS, the deliberate approach is to soften the link between the screening cues and any particular type of criminal victimization. In particular, for rape and sexual assault, as BJS translates

[5]This definition is different from that used by economists and other mathematical modelers, for whom "specification error" refers to an incorrect statement of an empirical model. We use the term differently in the report.

BOX 8-1
Definitions of Rape and Sexual Assault Used on the National Crime Victimization Survey

Rape—Forced sexual intercourse including both psychological coercion as well as physical force. Forced sexual intercourse means vaginal, anal, or oral penetration by the offender(s). This category also includes incidents where the penetration is from a foreign object such as a bottle. Includes attempted rapes, male as well as female victims, and both heterosexual and homosexual rape. Attempted rape includes verbal threats of rape.

Sexual Assault—A wide range of victimizations, separate from rape or attempted rape. These crimes include attacks or attempted attacks generally involving unwanted sexual contact between victim and offender. Sexual assaults may or may not involve force and include such things as grabbing or fondling. Sexual assault also includes verbal threats.

SOURCE: Bureau of Justice Statistics (n.d.-b).

these specific concepts into data collection, the respondent is asked the following question (Bureau of Justice Statistics, n.d.-d):

> *Has anyone attacked or threatened you in any of these ways:*
>
> - *With any weapon, for instance, a gun or knife,*
> - *With anything like a baseball bat, frying pan, scissors, or stick,*
> - *By something thrown, such as a rock or bottle,*
> - *Include any grabbing, punching or choking,*
> - ***Any rape, attempted rape or other type of sexual attack,*** *[emphasis added]*
> - *Any face to face threats,*
>
> *Or*
>
> - *Any attack or threat or use of force by anyone at all. Please mention it even if you are not certain it was a crime?*

The respondent also is asked a special follow-up question that focuses on how well the respondent knew the offender:

> *Incidents involving forced or unwanted sexual acts are often difficult to talk about. (Other than any incidents already men-*

tioned,) have you been forced or coerced to engage in unwanted sexual activity by:

(a) Someone you didn't know before
(b) A casual acquaintance
(c) Someone you know well?

The concept that BJS is trying to measure is clearly specified by the definitions it publishes. Unfortunately, these complex, multifaceted definitions are translated into a few simple words in the questionnaire: rape, attempted rape, other type of sexual attack, unwanted sexual activity. The cue screening approach may be effective for many common victimizations on the omnibus NCVS, but these words do not convey the complexity of the intended concepts nor capture the components of the BJS definitions of rape and sexual assault. (The section below on measurement error further discusses how respondents may misinterpret the words in these questions.)

CONCLUSION 8-5 There is serious specification error in the National Crime Victimization Survey measurement of rape and sexual assault. Although the Bureau of Justice Statistics has developed clear definitions of the concepts, they are replaced in the omnibus screener by ambiguous wording that does not convey the multifaceted concepts to respondents.

MEASUREMENT ERROR

Measurement error includes a large family of errors that may occur when response on a survey results in the collection of inaccurate or incomplete information. In this section, the report discusses potential measurement errors on the NCVS associated with the respondent, the questionnaire, the mode of collection, and with the interviewer/respondent interaction. These issues are interrelated, and each has the potential to result in measurement error on the NCVS.

Respondents

Survey research has mapped a respondent's cognitive process in answering survey questions (Schwarz, 1996; Strack and Martin, 1987; Tourangeau, 1984; Tourangeau, Rips, and Rasinski, 2000). In particular, Tourangeau, describes four steps a respondent goes through in responding to a survey question:

1. comprehending the question and instructions;
2. retrieving specific memories or information;
3. making judgments, that is, regarding the matching of the information to the question, and the completeness of that information; and
4. formulating a response.

In the final step of formulating a response, Cannell, Miller, and Oksenberg (1981) point out that the respondent evaluates potential responses based not only on whether he or she judges a potential response to be accurate, but also on other factors he or she views as important. This observation is important in assessing response to a sensitive question.

On the NCVS, a respondent may not comprehend a critical question in the same way that BJS intends.[6] For example, on the screening questionnaire (Bureau of Justice Statistics, n.d.-d), a respondent is asked:

Has anyone attacked or threatened you in any of these ways:

- *With any weapon, for instance, a gun or knife,*
- *With anything like a baseball bat, frying pan, scissors, or stick,*
- *By something thrown, such as a rock or bottle,*
- *Include any grabbing, punching or choking,*
- *Any rape, attempted rape or other type of sexual attack,*
- *Any face to face threats,*
  Or
- *Any attack or threat or use of force by anyone at all. Please mention it even if you are not certain it was a crime?*

In the "comprehension" process (see above), the respondent may focus on the listing of various weapons used to threaten and not understand that BJS would allow a *yes* response even if none of these types of threats were made. An example would be an incident of "date rape" involving alcohol use and being held down, but in which no weapon was used. Thus, the respondent might comprehend the question incorrectly, recall an incident from memory, make a judgment that the incident does not fit the criteria in the survey question, and respond *no*.

This screening question asks specifically about rape and attempted rape. The panel believes these terms are ambiguous for many respon-

[6]A similar example is discussed earlier in this chapter in the section on specification error: the panel has concluded that the concept of rape and sexual assault is misspecified in the data collection instruments of the NCVS. This section provides more discussion as to why a respondent may not understand these items in the way the BJS intends.

dents and many situations. This issue is discussed earlier in the chapter under "Specification Error" and is further discussed below in the section "Questionnaire."

Alternatively, the respondent may clearly understand the question, recall the memory of being raped at knifepoint, make a judgment that the incident is relevant to the question being asked, and yet decide not to disclose the incident. She or he formulates the *no* answer to the screening question based on the "other factors" (see Cannell, Miller, and Oksenberg, 1981; also see section on Item Nonresponse earlier in this chapter). As Rasinski (2012, p. 3) notes

> [W]hen the events are "sensitive" additional considerations for protecting the respondent's privacy, preserving the respondent's self-image and assuring the respondent that they will not suffer physical or psychological harm because of their disclosure must also be put into place.

See also Schaeffer (2000); Sudman and Bradburn (1982); and Tourangeau, Rips, and Rasinski (2000).

Questionnaire

The wording of questions is critically important to assist a respondent in comprehending the survey designer's intended meaning. Rasinski (2012) points out that in developing effective questions to solicit information about sexual victimizations, one must consider both the methodological aspects of designing sensitive questions and the specific nuances of talking about rape and sexual assault. There are several different aspects of a question that could make it sensitive—social undesirability, invasion of privacy, and risk of disclosure (to third parties) (Tourangeau, Rips, and Rasinski, 2000). A question that asks a respondent about experiencing sexual violence incorporates all three aspects of sensitivity.

Previous parts of this report discuss the issues of miscommunication when using such words as rape, force, and consent. Tracy et al. (2012, p. 3) explain this issue in a broad context:

> This historical context influences the way sex crime laws are written by lawmakers and enforced by law enforcement, and, in cases arising under those laws, how police decide whether to arrest, how prosecutors decide whether to take the cases to court, and how judges and juries make ultimate decisions as to whether to convict. The system's response in turn impacts whether victims perceive themselves as crime victims and whether they view the criminal justice system as one that recognizes them as crime victims. One consequence of the system's negative impact on victims is that it reduces victim reporting to and cooperation with police. Understanding this background will help in developing both survey instructions and

> questions that are more effective at capturing the prevalence and incidence of rape and sexual assault. It will also assist in understanding that the data collected may be limited to the extent that victim reporting—even to surveys—may be impeded by inaccurate societal notions about rape and sexual assault.

The NCVS is an omnibus crime survey, and the current screener uses cues that deliberately soften the link between the questions and any specific type of criminal victimization, and focuses instead on asking about such things as weapons and location. The panel has two major concerns about how the NCVS currently asks questions about rape and sexual assault. First, the questionnaire uses terms, such as rape, that do not always have consistent meaning and that do not clearly articulate the scope of actions that are included in the definition of rape. Second, the questions are embedded in a criminal context that may impede accurate reporting.

The NCVS uses the word rape, as in

> *Has anyone attacked or threatened you in any of these ways:*
> *(e) Any rape, attempted rape or other type of sexual assault*

The NCVS screener follows this with two other general cues regarding "incidents committed by someone you know" and "incidents involving forced or unwanted sexual acts." (See Chapter 4 for more details.) Taken together, these cues are meant to assist a respondent in recalling a rape or sexual assault.

As described in detail in Chapters 2, 3, and 4, the different legal statutes throughout the United States, the Federal Bureau of Investigation, and the BJS, each uses a different definition for the word "rape." It is not reasonable to assume that individual respondents will all interpret this word (or the term "sexual acts") identically. Important other surveys (described in Chapter 5) have taken a different approach. Although their questionnaires are not identical, they have used questions that more clearly describe specific "behaviors" that an offender may have exhibited. When a respondent is asked whether someone engaged in a very specific action in the incident, there is considerably less chance for miscommunication. These types of questions are referred to as "behaviorally specific" because they explicitly describe a set of behaviors. For example, on the National Violence Against Women Study, respondents were asked

> *Has a man or boy ever made you have sex by using force or threatening to harm you or someone close to you? Just so there is no mistake by sex, we mean putting a penis in your vagina.*

This question describes a specific action ("putting a penis in your vagina"), which is more likely to be clearly understood than asking a respondent if he or she has been raped. This approach was reinforced in a recent discussion of research methods for measuring rape and sexual assault (Jaquier, Johnson, and Fisher, 2011, p. 27):

> The usefulness of behaviorally specific questions cannot be overemphasized, not necessarily because they produce larger estimates of rape, but because they use words and phrases that describe to the respondent exactly what behavior is being measured. Using behaviorally specific screen questions appears to cue more women to recall their experiences.

Most of the studies that use behaviorally specific questions have measured a higher rate of incidence of sexual violence (Fisher, 2009), and it is the panel's judgment that the use of behaviorally specific questions improves communication with the respondent and facilitates more consistent responses.

CONCLUSION 8-6 Words, such as "rape" and "sexual assault," on the National Crime Victimization Survey may not be consistently understood by survey respondents. Other surveys have used more behaviorally specific words to describe a specific set of actions. More specific wording of questions would be understood more consistently by all respondents and thus lead to more complete and accurate answers.

The NCVS is a criminal victimization survey. It is introduced that way to household members. Once an interview begins, the questionnaire goes through a listing of crimes, asking each respondent if he or she has been the victim of any of them. When asked questions about rape and sexual assault, it is clear that the interviewer is asking about a crime. In fact, the questions about rape and sexual assault are embedded among questions that are dominated by other crimes. For example, as noted above, the following question is dominated by the descriptions of weapons and assaults.[7] Rape and sexual assault, particularly when no weapon is involved, may appear to be less central to the line of inquiry than other forms of assault in this list (Bureau of Justice Statistics, n.d.-d).

---

[7]The context and surrounding questions in a questionnaire may greatly affect responses on a survey. This was illustrated by Gibson et al. (1978, p. 251) in an experiment that added a series of attitude questions about crime to the National Crime Survey (NCS). They found that inclusion of the attitude supplement to the NCS had "a statistically significant and substantial impact on the victimization rates obtained."

Has anyone attacked or threatened you in any of these ways:

- *With any weapon, for instance, a gun or knife,*
- *With anything like a baseball bat, frying pan, scissors, or stick,*
- *By something thrown, such as a rock or bottle,*
- *Include any grabbing, punching or choking,*
- *Any rape, attempted rape or other type of sexual attack,*
- *Any face to face threats,*
  OR
- *Any attack or threat or use of force by anyone at all. Please mention it even if you are not certain it was a crime?*

Most sexual violence is committed by someone known to the victim. The victim may not have contacted the police (it is estimated that between 65 and 80 percent of such violent incidents are not reported to police) and may not think of the incident as a crime. The respondent may also think that because she or he did not contact the police about the incident, it should not be reported on a government crime inquiry. A respondent may fail to respond for these reasons even though the current NCVS screener has a cue reminding that "people often do not think of incidents committed by someone they know." Alternatively, the respondent may understand that the sexual victimization was criminal but may fear reprisal or may not want to get the other person "in trouble." Thus, the respondent may have reservations about answering questions about criminal incidents and the risk of disclosure to police.

CONCLUSION 8-7 Questions about incidents of rape and sexual assault in the National Crime Victimization Survey are asked in the context of a criminal victimization survey and embedded within individual questions that describe other types of crimes. This context may inhibit reporting of incidents that the respondent does not think of as criminal, did not report to the police, or does not want to report to police.

Data Collection Modes and Methods

Data collection mode can have important consequences for total survey quality. The mode affects the context of a survey: it affects questionnaire construction, the amount and type of communication with respondents, and the completion rate, among others. Considerable survey research regarding mode effects in surveys has been conducted. One of the most relevant, Tourangeau and Smith (1996), compared three methods (computer-assisted personal interviewing [CAPI], computer-assisted self-administered interviewing [CASI], and audio computer-assisted self-administered interviewing [ACASI]) of collecting survey data about sexual behaviors and other sensitive topics. Tourangeau and Smith (1996, p. 275) conclude

> The three mode groups did not differ in response rates, but the mode of data collection did affect the level of reporting of sensitive behaviors: both forms of self-administration tended to reduce the disparity between men and women in the number of sex partners reported. Self-administration, especially via ACASI, also increased the proportion of respondents admitting that they had used illicit drugs.

Thus a choice of data collection mode is very important when dealing with sensitive questions. A question may involve a potentially "socially undesirable" response. If an interviewer is asking the question, hearing the answer, and perhaps probing for more information, then the respondent may be concerned about the interviewer's approval or disapproval. Thus, a self-administered mode of collection generally provides respondents with less motivation to misreport on sensitive questions. In a review of reporting errors in surveys, Tourangeau and Yan (2007, p. 867) conclude

> [F]indings on mode difference in reporting of sensitive information clearly point a finger at the interviewer as a contributor to misreporting. It is not that the interviewer does anything wrong. What seems to make a difference is whether the respondent has to report his or her answers to another person.

The NCVS is interviewer administered. When the NCVS began, it relied more on in-person interviews with household members. This is still the method used for the first wave interviews. Beginning in 1980, cost considerations led BJS to use telephone interviewing (by the field representative) in subsequent waves, and telephone interviewing is now encouraged in all but wave 1. Approximately 57 percent of all within-unit interviews are conducted over the telephone. Because this percentage includes wave 1 interviews (which are primarily conducted in person) the percentage of telephone interviews for all subsequent waves is higher.

Yu, Stasny, and Lin (2008) reported a mode effect in the NCVS, with rape reported at a rate 1.45 times higher in personal interviews compared to telephone interviews. Using Bayesian methods, the authors estimated the probabilities that a personal crime that had occurred was not reported in the interview. "Thus for interviews conducted over the telephone with women who are victims of any type of personal crime (except for personal larceny), we estimate that approximately 37% of the women did not report the victimization" (Yu, Stasny, and Lin, 2008, p. 681). This analysis used unweighted data from the 1998 to 2004 NCVS for women respondents 16 years of age and older. (They also used 1993 to 1997 data as prior information in their Bayesian models.)

Privacy

The research findings on survey mode and asking sensitive questions raise a major concern with the current methods of data collection on the NCVS for measuring rape and sexual assault—a lack of privacy. As noted above, the NCVS is interviewer administered, with 43 percent of all interviews (including wave 1) conducted in person. The protocol involves a personal visit by the field representative to the selected address and an interview with each household member who is 12 years of age and older. The interviewing manual for field representatives on administering the NCVS states (U.S. Census Bureau and Bureau of Justice Statistics, 2008, p. A207):

> If *nonhousehold* [emphasis added] members are present, either in a sample housing unit or a group quarters, ask the respondent if he/she wishes to be interviewed in private. If so, make the necessary arrangements to either interview the person elsewhere or at a different time. Some respondents may prefer not to be interviewed while other household members are present. Always respect the respondent's need for a private interview.

Thus, the interviewer manual indicates that some respondents may prefer a private interview but does not direct the field representative to ask unless nonhousehold members are present. The training material used in the refresher training in 2011 did not cover the need for privacy during individual interviews (U.S. Census Bureau and Bureau of Justice Statistics, 2011a, 2011b).

The panel believes that privacy in interviewing about sexual violence is critical because most rapes and sexual assaults are committed by individuals whom the victim knows. The offender may, in fact, be member of the household. Another possibility is that a teenager has been a victim of date rape but has not told his or her parents. A respondent who has been sexually victimized may not report the victimization if that reporting may be overheard or otherwise inferred by another household member. This concern goes beyond whether there is another household member in the same room during the interview, to the situation in which the interview can be overheard from another room in the home, to the situation in which another household member may notice that the victim's interview lasted longer than the one in which he or she participated. As Tourangeau and Yan (2007, p. 862) conclude, "respondents may be reluctant to report sensitive information in surveys partly because they are worried that the information may be accessible to third parties," as outlined above. Other researchers have concluded that the effect of the presence of others when responding to sensitive questions is dependent on whether the bystander already knows the information that is being requested (Aquilino, Wright, and Supple, 2000).

Tourangeau and Yan (2007) reviewed research on the effect of the presence of others in reporting on sensitive questions. The results were mixed and very situational. They found that a spouse's presence did not appear to have a significant overall effect on survey responses, but they found a highly significant effect of parental presence, which reduced socially undesirable responses.

Yu, Stasny, and Li (2008) found that the presence of a spouse during an NCVS interview likely led to the underreporting of incidents of rape and sexual assault. The authors used data from the 1998 to 2004 NCVS for women respondents 16 years of age and older. (They also used 1993 to 1997 data as prior information in their modeling.) They categorized personal interviews by "who was present" during the interview, coded by the field representative: (i) spouse and no one else, (ii) spouse and at least one other person, (iii) at least one person but not the spouse, and (iv) no one else present. Telephone interviews were categorized as "unknown" because the field representative did not know who might be present on the other end of the phone line. In an analysis of unweighted data, Yu, Stasny and Lin (2008, p. 671) found that "compared with a woman who was interviewed alone, rape (including rape, attempted rape, and sexual assault) was reported about one-fifth as frequently when a spouse was present." As discussed in an earlier section of this report, they also reported a mode effect, with rape reported at a rate 1.45 times higher in personal interviews compared to telephone interviews. They referred to a telephone interview or the presence of the spouse in a personal interview as a "gag factor" (Yu, Stasny, and Lin, 2008, p. 666). Using Bayesian methods, the authors estimated the probabilities that a crime was not reported in the interview. "Thus for interviews with women who are victims of rape and whose spouse was present during the interview, we estimate that 86% of the women did not report the victimization" (Yu, Stasny, and Lin, 2008, p. 681).

Several factors make privacy an elusive goal in the NCVS data collection. First, a dwelling may not have a private location where other household members neither see nor hear what is going on. Second, rape and sexual assault are two relatively low-incident criminal victimizations among the many more victimizations that the NCVS measures. Most of the other victimizations involve less sensitive questions, and the field representative's main goal is to get a completed questionnaire from each household member. The training for interviewers does not stress the need for privacy, and the field representative is likely to view the need to have a completely private conversation as secondary to getting the completed interviews. Third, each household member (12 years of age and older) is interviewed and therefore knows what the others are being asked. This fact, in itself, may cause a victim to feel intimidated when asked to disclose experiences of sexual violence.

Telephone interviewing of household members may offer some privacy. Because field representatives usually make these telephone calls from their own homes, they are directed to make sure that their own family members or neighbors cannot listen to the telephone call. Again, the interviewer manual does not direct the field representative to ask the respondent to try to find a private location in the home (U.S. Census Bureau and Bureau of Justice Statistics, 2008). Moreover, there may not be such an area where other household members cannot hear the respondent's side of the conversation. The NCVS requires the respondent to describe events in the incident report, which might be overheard by other household members. A respondent may also have concern that another household member may try to listen on an extension. The research on this issue is well summarized by Tourangeau and Yan (2007, p. 867):

> [The] findings on this issue [telephone interviewing of sensitive questions] are not completely clear, but taken together, they indicate that the interviewer's physical presence is not the important factor.... On the whole, the weight of the evidence suggests that the telephone interviews yield less candid reporting of sensitive information.

CONCLUSION 8-8 The current data collection mode and methods of the National Crime Victimization Survey do not provide adequate privacy for collecting information on rape and sexual assault. This lack of privacy may be a major reason for underreporting of such incidents.

Interviewer-Respondent Interactions

The NCVS is an interviewer-administered survey. As such, the interaction between the interviewer and the respondent during the interview heavily influences the quality of survey responses. In this section, the report looks at issues associated with interviewers, including gender, training and preparation, and monitoring.

Gender

As discussed above, the presence of an interviewer may lead to misreporting on certain sensitive questions if the respondent is reluctant to talk about socially undesirable opinions or incidents. If an interviewer is administering the survey, then the gender of the interviewer may also influence (either positively or negatively) a respondent who is asked a sensitive question. Catania (1997) found higher item-level response rates to questions regarding same-sex sexual experiences to interviewers of their own gender. Another study examined whether the gender of the voice in ACASI might affect responses to a set of sensitive questions asked of young adults. The finding suggests that female interviewers may get more accurate reports (Dykema et al., 2012, p. 312):

> [There were] higher levels of engagement in the behaviors and more consistent reporting among males when responding to a female voice, indicating that males were potentially more accurate when reporting to the female voice. Reports by females were not influenced by the voice's gender.

A standard operational practice on surveys of sexual conduct or violence has been to use female interviewers. Female interviewers were used exclusively in the National Women's Study, the National College Women Sexual Victimization Study, and the National Intimate Partner and Sexual Violence Survey (discussed in Chapter 5). The National Violence Against Women Study (discussed in Chapter 5) incorporated a test of interviewer gender, using female interviewers for female respondents, and using both male and female interviews for male respondents (Tjaden and Thoennes, 2000). The NCVS uses mostly, but not exclusively, female interviewers.[8] The panel agrees with this standard practice, but believes that additional research is needed for definitive answers regarding the effect of an interviewer's gender separate from other factors. Survey organizations are increasingly coding the demographic characteristics of interviewers (such as gender and age) that might affect recruiting and response quality so that possible effects can be more thoroughly studied. The results from these efforts will be important for the design of all surveys on sensitive topics.

Training and Preparation

Interviewers need to receive high-quality training to reduce interviewer effects and deliver survey responses of high quality. The Census Bureau understands this important aspect of the survey process and strives to train its field representatives appropriately for these complex surveys. However, there are two issues with the training provided to interviewers on the NCVS: overall training and the rarity of the incidents of interest.

The first issue is that the overall training effort on the NCVS has been inadequate. Refresher training of interviewers on the NCVS was eliminated during a 10-year period due to budget restrictions. The agencies acknowledged the problem (U.S. Census Bureau and Bureau of Justice Statistics, 2011a, pp. 1-5):

[8]The Census Bureau faces issues related to equal employment opportunities when considering hiring based on gender.

> [G]eneral performance reviews and refresher training were eliminated. So while the survey remained in the field and we were still able to generate annual crime estimates, we (Census) and the BJS had limited ability to monitor the quality of the data collected and to ensure that our field staff fully understood what was expected of them.

Fortunately, some training is being restored (Bureau of Justice Statistics, 2012a, p. 11):

> Beginning in August 2011, refresher training of all field representatives (FR) was conducted using an experimental split sample cluster design. This was the first comprehensive refresher training that had been conducted since the 1990s. To maintain consistent year-to-year comparisons, Census and BJS implemented the experiment in a manner that isolated the effects of training without contaminating the annual 2011 estimates.

An issue with the training of interviewers is that rape and sexual assault is only one type of victimization among many on the NCVS questionnaire, and it is rarely reported. However, questions that ask about this topic require special sensitivity from interviewers. The NCVS refresher training for field representatives was 1.5 days in length, during which the NCVS screener was discussed for only 2 hours. Moreover, that 2 hours included not only discussion of methods of asking sensitive screener questions but also many other issues. The trainers provided a number of useful suggestions to follow when interviewing victims of sexual or other sensitive crimes (see Box 8-2). The panel applauds this refresher training, which covered many facets of the NCVS. However the limited time devoted to training on asking sensitive questions, the need for privacy in asking those sensitive questions, and a fuller understanding of sexual victimization did not get the emphasis that is needed in order to ensure complete reporting. And even if adequate training could be provided, such training would not be reinforced through the day-to-day survey process because the NCVS is a general-purpose criminal victimization survey, and an interviewer very infrequently gets a positive response on questions about rape and sexual assault.

CONCLUSION 8-9 The current training for National Crime Victimization Survey interviewers with regard to the subject of rape and sexual assault is insufficient to ensure complete and accurate responses. Moreover, because interviewers only infrequently encounter reports of these crimes, they do not get the opportunity to practice and reinforce the training that they do receive.

BOX 8-2
Suggestions When Interviewing Victims of Sexual or Other Sensitive Crimes Provided to National Crime Victimization Survey (NCVS) Interviewers During Refresher Training in 2011

- Be sensitive to what the respondent is telling you; however, keep the respondent on track because some respondents may have the tendency to tell you more than what is being asked.
- Be respectful and polite to victims, even to those who do not want to talk.
- Avoid unnecessary pressure. Be patient.
- Be supportive and let victims express their emotions, which may include crying or angry outbursts.
- Be careful not to appear overprotective or patronizing.
- Avoid judging victims or personally commenting on the situation.
- Remind the respondents periodically, if necessary, about the importance of their responses.
- Reassure the respondent that knowing the prevalence of the form of violence they are experiencing will be useful to expand efforts to identify ways to help victims of that type of crime and to hold perpetrators accountable.
- Supply the respondent with a copy of the NCVS-110 Fact Sheet brochure (show example), which contains several hotline numbers that they may find helpful to call if the person asks for assistance. Make sure you have an ample supply of the Fact Sheet to provide to respondents when needed.

SOURCE: U.S. Census Bureau and Bureau of Justice Statistics (2011a, pp. 2-33-2-34).

Monitoring

Monitoring of interviews is a method of ensuring quality control over the interview process, improving interviewer performance and improving data quality. It is a standard practice for central location telephone interviews. It is more limited with field interviewing and with decentralized telephone interviewing. Thissen et al. (2009, p. 2) provide an overview of the classic techniques of monitoring in field data collection and their advantages and disadvantages. These techniques include in-person observation; post-interview discussions with interviewers; verification contact, by telephone or in-person reinterview; review of response data and timers; and tape recording during interview.

The NCVS is collected with a combination of field and telephone interviews conducted by the field interviewers. Thus, there is currently no centralized telephoning that is monitored on a continuing basis. Previously, a periodic NCVS quality control recontact had been conducted as part of interviewer evaluations, but this process was suspended over several years (along with refresher training) because of budget constraints. In this process, the reinterviewer verified several things (U.S. Census Bureau and Bureau of Justice Statistics, 2010):

- the correct sample units were interviewed,
- the listing sheets were completed or updated properly,
- the household screens were completed or updated properly,
- all screen questions were asked and all answers recorded, and
- any noninterviews were classified accurately.

The research on interviewer monitoring came mostly from centralized telephone interviewing because field interviewing relied almost exclusively on a "verification contact" until the introduction of computer audio-recorded interviewing (CARI) in 1989.[9] CARI is a laptop computer software application that unobtrusively digitally records the audio exchange between an interviewer and a respondent during interviews. The software is programmed so that individual questions or sections will automatically be recorded for quality review. After the interview is completed, the audio files are downloaded and transmitted to the central program staff for coding and review.

The CARI technology not only records interviewer-respondent verbal interactions but also ensures that the description of the interview is not biased:

1. It records unobtrusively because the microphone is built into the computer;
2. The microphone is activated at the appropriate points by the computer program not the interviewer, which not only reduces intrusiveness but also makes the recording independent of the interviewer; and
3. The digital recordings are exported as audio files of individual questions that can be sorted by question, respondent, or interviewer, which permits rapid and efficient purposive review.

In a feasibility report, Biemer et al. (2000, p. 1) identified the range of applications, including

[9]The method was developed and pioneered by RTI. It was first deployed in a national field study in the 1989 National Survey of Child and Adolescent Well-Being.

- detecting gross departures from appropriate procedures, including interview fabrication;
- evaluating interviewer execution of interviewing guidelines, which permits corrective feedback for future interviews as well data quality control for existing interviews;
- identifying questionnaire problems and data collection difficulties using interviewer-respondent interaction coding; and
- collecting verbatim responses to open-ended questions in an interview.

These applications include all of the goals of monitoring within a centralized telephone-interviewing facility with the exception of immediate feedback.

The researchers (Biemer et al., 2000) found that the CARI-based approach was less expensive than other traditional approaches for interview verification of field interviews: 23 percent less expensive than face-to-face follow-up and 32 percent less expensive than a telephone- and postcard-based approach. However, these analyses ignored the system development costs of a functional CARI system.

It is particularly noteworthy that the CARI system was piloted on an extremely sensitive survey, the National Survey of Child and Adolescent Well-Being, which is a panel survey of 6,700 children who are the subjects of reports of abuse and neglect. The study required both signed and audio-recorded consent to use CARI. Consent to use CARI was obtained in 85 percent of the caseworker interviews, 83 percent of the caregiver interviews, and 82 percent of the child interviews.

CONCLUSION 8-10 Monitoring of interviewers is important to ensure quality and to identify areas in which an individual interviewer needs reinforcement and areas in which improved training is needed. The monitoring method used in the National Crime Victimization Survey, periodic reinterviews of selected respondents, is not adequate to ensure interviewing quality.

9

Synopsis of Potential Errors in the National Crime Victimization Survey

The mission of the Bureau of Justice Statistics (BJS) is "to collect, analyze, publish, and disseminate information on crime, criminal offenders, victims of crime, and the operation of justice systems at all levels of government" (Bureau of Justice Statistics, n.d.-c). A major tool in accomplishing this mission is the National Crime Victimization Survey (NCVS), which provides a rich source of information on criminal victimizations, the victims, and the specifics of their criminal victimizations. In a 2008 report, the National Research Council found that the NCVS provided the only annual source of national data on criminal victimization with detail at the incident level. The report's first recommendation (3-1) was that "BJS must ensure that the nation has quality annual estimates of levels and changes in criminal victimization" (National Research Council, 2008, p. 3). This panel strongly concurs with that recommendation. To further that mission, this panel has focused specifically on the quality of the measurements of rape and sexual assault with the purpose of helping improve quality in BJS's estimates of those types of criminal victimizations.

The NCVS has both sampling and nonsampling error, as does every survey. The panel examined the error structure of the NCVS (see Chapters 7 and 8) in considerable detail with the intent of identifying areas that might be particular problems. We reiterate that the NCVS is an omnibus survey with the primary goal of estimating many types of criminal victimization. Its survey design is geared toward this broad goal, and it does not and cannot incorporate separate features for measuring each different type of victimization. Yet the survey's basic goal and design create problems for measuring rape and sexual assault. This chapter summarizes the discussions in Chapters 7 and 8 that the panel believes are most problematic for measuring rape and sexual assault. The chapter concludes with several specific recommendations. The bulk of the panel's recommendations are included in Chapter 10.

POTENTIAL SOURCES OF ERROR: SUMMARY

The NCVS uses a classical area sampling design based on selection in multiple stages to create a sample of housing units. This basic design is used for a number of general household surveys conducted by the Census Bureau, and it is a good design for an omnibus type survey. For estimates of rape and sexual assault, however, a limitation of the NCVS sample design is its current inability to control the sampling error for estimates of these low-incidence events. Although the sample size of the NCVS is large, rape and sexual assault are low-incidence events and the size of the estimated coefficients of variation (CVs)[1] for those estimates at the national level allows considerable year-to-year fluctuation (see Chapter 7). In addition, also because of the low incidence of rape and sexual assault, BJS has to pool multiple years of data to get estimates of rape and sexual assault for important subpopulations (e.g., by race, gender, marital status, or age group).

In 2011, BJS changed its process for counting series victimizations. The effect of the change was substantial for estimates of rape and sexual assault: these victimizations were undercounted in the past, but series victimizations (based on only a few reports) now account for almost 40 percent of the national estimates of rape and sexual assault. The handling of these outlier reports has created additional concern about year-to-year fluctuation in the estimates of rape and sexual assault.

The overall response rates on the NCVS have remained fairly high, and there is little item nonresponse coded for specific rape and sexual assault screener questions. However, the panel is concerned about panel attrition, particularly because those individuals most likely to drop out—younger people and individuals not living as a couple—may be at greater risk for rape and sexual assault. The panel's judgment is that individuals who respond on the NCVS but do not want to report a specific victimization will refuse the question indirectly by responding that no victimization occurred. This item nonresponse would then be coded as a legitimate zero.

Although BJS clearly defines what it intends to measure regarding rape and sexual assault, because of the omnibus design of the screener, these concepts are poorly translated through the data collection instruments. The in-

[1]The CV is the standard error of a survey estimate divided by the estimate itself (expressed as a percent). It provides a relative measure of the sampling error associated with survey estimates.

struments use oversimplified terms that can be interpreted in different ways. This situation creates the potential for specification error in the NCVS.

Several issues with the NCVS could lead to measurement error associated with the questionnaire, the data collection mode, and the interaction between interviewers and respondents. These issues include comprehension of key questions and respondents' willingness to answer certain questions. The latter could be affected by the mode of data collection and interviewer gender. There continues to be inadequate training of interviewers and monitoring of the interviews on the NCVS.

All surveys have error associated with their processing. BJS publishes little information about its processing methods or errors found. The lack of such transparency makes it difficult for NCVS data users to fully understand NCVS estimates and their limitations.

OBSTACLES TO HIGH-QUALITY ESTIMATES

Based on the potential errors in the NCVS discussed in Chapters 7 and 8 and summarized above, the panel identified four major obstacles for accurately estimating incidences of rape and sexual assault.

1. a sample design that is inefficient for measuring these low-incidence events,
2. the context of "crime" that defines the survey,
3. a lack of privacy for respondents in completing the survey, and
4. the use of words with ambiguous meaning for key measures in the questionnaire.

These obstacles form the basis for the panel recommendations in Chapter 10 to create a separate survey to measure rape and sexual assault.

Inefficiency of the Sample Design

The NCVS is a general criminal victimization survey that targets the noninstitutionalized population of the United States with questions about different types of criminal victimizations, including crimes against property and people. Of the 19 million criminal victimizations identified through the NCVS in 2011, 64 percent were property crimes, which are dominated by theft; 29 percent were violent crimes (against people) but not classified as serious; and the remaining 7 percent of criminal victimizations were classified as serious violent crimes,[2] of which only 1 of the percentage points

[2]Serious violent crimes include robbery and aggravated assault, but not murder and kidnapping, which are not estimated through the NCVS.

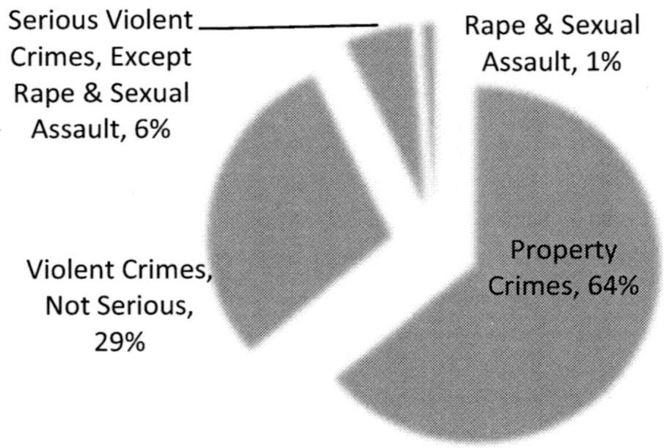

FIGURE 9-1 Criminal victimizations by type, NCVS 2011.
SOURCE: Data from *Criminal Victimization 2011* (Bureau of Justice Statistics, 2012a).

was rape and sexual assault (see Figure 9-1). Thus, from a statistical perspective, rape and sexual assaults are statistically rare incidents.

The rarity of an attribute in a population presents unique challenges to efficient and effective sampling to estimate the proportion of the population with that attribute. In this situation, the difficulty can be seen by observing the CVs for various types of criminal victimizations, with a lower number meaning a more precise estimate (see footnote 1). Nationally, in 2011, the CV for property crime was 2 percent, for serious violent crime it was 7 percent, and for rape and sexual assault it was 14 percent (Bureau of Justice Statistics, 2012a). The NCVS sample design is not well suited to measure the uncommon attribute of having experienced a rape or sexual assault in the past 6 months (see Conclusion 7-1 in Chapter 7).

This concern is intrinsic to the basic design of the NCVS. BJS and the Census Bureau could address this issue by implementing a very large increase to the existing sample size of the NCVS or by modifying the NCVS sample design to focus more efficiently on subpopulations at greater risk for sexual victimization. (See Chapter 10 for a discussion of such a focused design.) However, the NCVS is an omnibus victimization survey whose goal is to measure a broad range of criminal victimizations. If it were to focus on the subpopulation at risk for a specific type of victimization (rape and sexual assault), then it would likely make the survey design much less efficient for measuring other types of criminal victimizations across the whole survey population.

Criminal Context of the Survey

The NCVS is a criminal victimization survey. It is introduced that way to household members, and the questionnaire goes through a listing of different types of crimes, asking each respondent whether any of these crimes has happened to him or her. As discussed in Chapter 8, victims may not always think of a rape or sexual assault as a crime, particularly if the respondent knows the offender. In addition, some victims may fear disclosure to police and may associate a government crime survey too closely with law enforcement. The panel concludes that the context of a criminal victimization survey is a major barrier to accurate reporting of incidents of sexual victimizations (see Conclusion 8-7 in Chapter 8).

BJS could address this issue by moving the measurement of rape and sexual assault to a separate survey (as recommended in Chapter 10) or by changing the entire context of the NCVS to something more neutral. The modification of the NCVS into a survey that has a "neutral context" would entail a major redesign of that survey (including its name), and such a change would undermine other goals of the survey. The context of criminal victimization appears to work well for reporting other, less sensitive, victimizations. A redesign of the NCVS to remove the "criminal context" on this large omnibus survey would make it difficult to effectively communicate the purpose of the survey to respondents, and it would likely negatively affect the reporting of many other types of victimizations.

Lack of Privacy in Responding

Providing a respondent with privacy is an important prerequisite for any survey that deals with sensitive questions. Privacy, specifically from other household members, is critical for accurately responding to inquiries about rape and sexual assault, in part because the victim often knows the offender. In fact, the offender may be a household member. The current NCVS data collection protocols do not provide sufficient privacy (see Conclusion 8-8 in Chapter 8). Some privacy gains would be possible on the NCVS by switching to a self-administered mode of data collection. This switch would be a major change for the survey, and it would likely have both beneficial and detrimental effects on collecting information about other types of criminal victimizations.

The panel concludes that such a change of data collection mode is important to make, but that this change alone would still not provide the level of privacy that is needed for reporting sexual victimizations. As long as multiple members of the household are interviewed with an identical questionnaire, an adequate level of privacy is unlikely to be obtained. Each household member knows the questions that are asked on the survey and the duration of the interview based upon their own responses. An individual household respondent may continue to have concerns about reporting an incident of rape or sexual assault under those circumstances. To achieve the privacy level needed for questions regarding rape and sexual assault, the survey protocol would also need to change to targeting only a single person in each household or to deploy a "matrix" design in which each household member would be screened for different types of criminal victimizations. If implemented for the NCVS as a whole, then either approach would severely reduce the number of responses obtained for each specific type of victimization. Without an accompanying increase in sample size, this change would negatively affect the precision for estimates of all types of victimizations.

Ambiguous Terms in the Questionnaire

The current NCVS screening questionnaire uses a series of cues to aid respondents in remembering past victimizations. As part of those cues, it uses the terms *rape* and *sexual assault* without any further elaboration as to what those terms mean. Although BJS has precise meanings for these terms, they are not provided to respondents. It is not reasonable to assume that all individual respondents will interpret these words identically or as BJS anticipates (see Conclusion 8-6 in Chapter 8). More behaviorally specific words that describe a specific set of actions would lead to more accurate responses on these questions.

Such a behavioral approach is not completely alien to the NCVS. Some of the screening questions for *assault* do use some behaviorally specific language: a respondent is not asked whether he or she has been assaulted, but is asked questions that describe specific behaviors—"*Has anyone attacked or threatened you . . . with anything like a baseball bat, frying pan, scissors, or stick?*" It would be possible to make behaviorally specific changes on the current NCVS regarding rape and sexual assault with only a modest effect on other parts of the questionnaire. However, there would need to be specific screening for rape and sexual assault that is separate from and in addition to the more generalized cuing sequences currently used in the screener. Another difficulty is that the NCVS interviews children as young as 12 years of age, and there may be parental objections to the behaviorally specific questions. With a separate survey as recommended by the panel, there would be more options available to BJS on handling respondents between the ages of 12 and 17.

GENERAL RECOMMENDATIONS

The majority of the panel's recommendations specifically address new directions for measuring rape and sexual assault (see Chapter 10). In this section, the panel offers several general recommendations to BJS regarding improved documentation and future research that grew out of the panel's review of the NCVS. With regard to research, we consider the unaddressed topic of child victimization.

The error review contained in Chapters 7 and 8 and summarized above identifies potential errors on the NCVS. The panel did not have the time or resources to conduct a complete error profile of the NCVS, which would include measuring the levels of error and their effects on the estimation of rape and sexual assault. The panel recommends this action to BJS.[3]

RECOMMENDATION 9-1 The Bureau of Justice Statistics should conduct an in-depth total error profile of the National Crime Victimization Survey, specifically focusing on estimation of rape and sexual assault. This profile should be made available to public data users.

Throughout its work, the panel found that the publicly available documentation was sometimes difficult to locate on the BJS website, and the major methodology document was several years out of date. The age and lack of clarity of existing documentation of the NCVS inhibits a complete and accurate understanding of the NCVS methodology.

RECOMMENDATION 9-2 The Bureau of Justice Statistics should provide complete documentation of the methodology in its surveys, details of survey response, costs, and the individual components of total survey error. The documentation should be made publicly available and easy to access.

Although BJS has commissioned a number of research studies over the past 10 years and many of the final reports of those studies are on the BJS website, they are inadequately organized to allow data users easy access. An interested data user has to know that a particular research study was conducted and then do a "hit or miss" search to see whether a research report on that topic was posted. In addition, the document *Survey Methodology for Criminal Victimization in the United States* (Bureau of Justice Statistics, 2008b) has been periodically updated, but the most recent update appears to be 2008.

---

[3]To provide guidance in implementing this recommendation, here are some examples of error profiles in the literature: Brooks and Bailar (1978); Chakrabarty and Torres (1996); Doyle and Clark (2001); Jabine, King, and Petroni (1990); Kalton et al. (2000); and U.S. Energy Information Administration, 1996. Kasprzyk and Kalton (2001) review the use of quality profiles in U.S. statistical agencies and discuss their strengths and weaknesses for survey improvement purposes.

RECOMMENDATION 9-3 The Bureau of Justice Statistics should clearly lay out its research agenda on the agency's website and link the agenda topics to completed reports. For research in progress, the website should include information about what is being done and when a report is expected.

The NCVS and this report do not address the victimization of children under the age of 12 years, a serious and neglected problem in the overall measurement of sexual victimization. There is a partial accounting of the victimization of children because several of the public health–oriented surveys that measure lifetime prevalence of rape and sexual assault have asked respondents their age when first victimized. For example, the National Intimate Partner and Sexual Violence Survey estimated that of individuals who reported having been raped in their lifetime, 12.3 percent of females and 27.8 percent of males were first raped before the age of 12 (Black et al., 2011).

The panel believes that the victimization of young children is a serious problem that requires further attention. No ongoing survey captures this information, with child protective services providing the major source of information on these victimizations. Unfortunately, this situation is similar to when the Uniform Crime Reports was the single source reporting on the sexual victimization of adults, with resulting concerns about underreporting to authorities. However, the issues associated with interviewing children mean that the measurement of this type of victimization needs a specialized approach—it cannot be measured through the same vehicle and the same methods that are used for adults and older children. The panel did not have the time and resources to pursue this extensive area of research. Instead, it recommends that BJS explore options to measure these sexual victimizations in the future. This might include the inclusion of retrospective questions to adults about childhood victimizations.

RECOMMENDATION 9-4: The Bureau of Justice Statistics should begin research to explore options for measuring the incidence of rape and sexual assault of children younger than 12.

10

New Directions for Measuring Rape and Sexual Assault

The panel was charged with recommending best methods for obtaining survey statistics on rape and sexual assault on household surveys of the Bureau of Justice Statistics (BJS). In reviewing all of the material presented earlier in this report, the panel thinks that it is highly likely that the National Crime Victimization Survey (NCVS) is underestimating rape and sexual assault. The panel, with limited resources, was not able to measure the extent of such an undercount with statistical rigor.

The previous chapter discusses four major obstacles to quality on the current NCVS: (1) a sample design that is inefficient for measuring these low-incidence events, (2) the context of "crime" that defines the survey, (3) a lack of privacy for respondents in completing the survey, and (4) the use of words with ambiguous meaning for key measures in the questionnaire.

As detailed in Chapter 9, only one of the four obstacles, the last one—use of ambiguous terms—can be readily addressed within the structure and operations of the current NCVS without negatively impacting the estimation of other important types of criminal victimizations. These obstacles led to the conclusion that the needed changes could not be adequately implemented within the framework of the existing survey, and that best practices for measuring rape and sexual assault would require that these measurements be decoupled from the NCVS.

CONCLUSION 10-1 The best methods for measuring rape and sexual assault cannot be implemented without separating that measurement from the measurement of other criminal victimizations.

RECOMMENDATION 10-1 The Bureau of Justice Statistics should develop an independent survey—separate from the National Crime Victimization Survey—for measuring rape and sexual assault.

In the rest of this chapter, the panel presents its recommendations for developing the recommended new survey for estimating rape and sexual assault, for important features of that survey, and for research needed as part of that development. As a part of the entire program design, the panel also offers several recommendations regarding improved training and monitoring of interviewers and regarding obtaining useful input and feedback from data users and the broader statistical community.

A NEW SURVEY TO MEASURE RAPE AND SEXUAL ASSAULT

Sample Design

Enhanced Multiple-Frame Design

The proportion of a population with a specific attribute (in this case, having been victimized by rape or sexual assault) can be estimated with greater precision by isolating population subgroups with relatively higher attribute rates and then sampling those subgroups more intensively than the rest of the target population. The higher the attribute rate in a subgroup, the greater potential gains in precision. The first challenge in this approach is to identify subgroups of people who are at higher risk of rape and sexual assault criminal victimizations than the general population. A second challenge is to identify such groups in a way that does not isolate previous victims or make them fearful or reintroduce the trauma of the incident.

In spite of the difficulty, the introduction of one or more additional sampling frames derived from sources likely to include persons victimized by rape and sexual assault will be a more cost-efficient alternative to the current NCVS frame. The panel strongly urges BJS to explore these opportunities.

RECOMMENDATION 10-2 The Bureau of Justice Statistics should rigorously compare the relative cost-efficiency of alternative sample designs for the recommended new survey to measure rape and sexual assault, including the multiple-frame approach described in this chapter.

An enhanced design would be anchored by a frame with coverage of the general population, such as the current NCVS frame. An initial step toward this enhanced design would be to more effectively target stratification in selecting the general-purpose sample to oversample areas where the income and demographics characteristics would be likely to make household members more at risk for sexual victimization. A second step would be to sample the group quarters frame with relatively higher rates because many types of group quarters may include relatively high concentrations of victims of rape and sexual assault (e.g., college residence halls, group shelters, etc.).

This general-purpose frame would be enhanced with one or more additional frames that would focus on specific subgroups that are at higher risk for sexual assault but are broader than a list of known sexual assault victims. Sources that might be used as a sample frame include, but are not limited to

- female college students,
- women who use Indian health service facilities,
- assault cases known to law enforcement,
- people treated for trauma in hospital emergency rooms,
- people who have filed a police report for any type of serious violent crime,
- residents of shelters for abused and battered women, and
- outpatients from mental health clinics.

These sources are not lists of rape and sexual assault victims but would include a much higher percentage of victims than seen in the general population (providing a statistical advantage to using them). Specialized sample selection and recruitment strategies may need to be developed to minimize concerns from victims who are found through supplemental frame sources. Following those procedures, the sample of individuals would be interviewed using the same survey instrument regardless of the frame from which they were selected. Note that a respondent's victimization status (for rape and sexual assault) would be unknown when an interview begins: it would be determined solely from answers given on the questionnaire. However, because of the addition of these specialized frames, the resulting sample is likely to have more positive responses to the questions on sexual victimization. This will make the estimates more precise, while the survey weights will ensure that this process does not artificially inflate the criminal victimization rate.

This approach has significant drawbacks. First, it will be very difficult to get access to some of the administrative sources for the supplemental frames. For example, police reports are not available to the general public, but they may be available in some form to a government agency. Second, HIPAA regulations[1] may restrict access. Third, there is concern that a victim who has experienced sexual abuse may not want his or her name shared

[1]HIPAA is the federal Health Insurance Portability and Accountability Act of 1996.

with a government survey group, and a person who has sought help from a hospital, shelter, or health clinic may feel betrayed if that institution provides names to BJS. These are serious concerns that will need to be worked through, but there is prior experience in the survey field working with respondent recruitment from these types of sources.

A fourth drawback is that working with more frames and the overlap between frames would add complexity and costs for frame development, sampling, and recruitment (Hartley, 1962; Lohr, 2011). In addition, because one cannot know in advance how well this approach would work, there is a basic question of whether the precision gains from the design will outweigh the added complexity. The exact answer to this question can only be determined through specific research. However, the panel did a preliminary cost-efficiency analysis that was very encouraging; it merits a thorough examination by BJS.

The panel's analysis looked at a simplified case of two overlapping sampling frames with simple random sampling in each frame.[2] One frame (the administrative frame) is completely contained in the other more complete frame (the household frame). For example, individuals who were treated for trauma in an emergency room are also a part of the U.S. population. For simplification in this illustration, overlap between the two frames is handled by screening out members of the administrative frame from household members in the sample chosen from the household frame. Under this simplified case, the panel observed the relative precision of a cost-variance optimized dual-frame estimator (in comparison with a single-household frame) of the proportion of the population with the attribute. In a specific example, the dual-frame design had a variance that was 43-45 percent lower than the cost-equivalent variance for the single-household frame design (see Box 10-1 and more detail in Appendix E).

The panel's analysis illustrates that the added cost of data collection on the administrative frame is not an important factor in this scenario. What does matter is the proportion of the population in the administrative frame that has the attribute (see Table 10-1). As the table shows, the relative efficiency of the dual frame decreases significantly as that proportion decreases. The variance reduction shown in this table would be somewhat offset with a more complex formula that takes into account the cluster effects of the sample.

Research is needed to identify specific administrative frames for which there is a higher risk of rape and sexual assault that may be used in this enhanced multiple-frame design without infringing on the privacy of victims. Research is also needed to further assess the effects of the cluster

[2]Details of the analysis can be found in Appendix E and in a paper presented at the recent Joint Statistical Meetings (Kalsbeek, Spencer, and House, 2013).

BOX 10-1
Comparing Precision from Dual- and Single-Frame Alternatives

In this scenario, police records were to be used to define an administrative frame of rape victims, of size $N_A \sim 140{,}000$ (extrapolated) to the total U.S. population from 1997 data from the Uniform Crime Reports (UCR) summary. Based on National Crime Victimization Survey (NCVS) data (average per 1992-2000), there were 116,300 such criminal victimizations reported to police. Thus the proportion of the population within this administrative frame, experiencing this criminal victimization, P_A, is estimated as:

$$P_A = 116{,}300 / 140{,}000 = 0.83$$

The household frame (similar to that used by the NCVS) would be much larger, N = 250,000,000. The population proportion of the same attribute (based on *Criminal Victimization, 2007*) is much smaller,

$$P = 0.001$$

Costs for the NCVS (Rand, 2009) for fiscal year 2009 were $26 million to collect approximately 150,000 interviews. Thus the cost per completed interview,

$$C_{HH} = \$26 \text{ million} / 150{,}000 = \$173$$

Results for fixed total cost:

If the *cost of data collection using the administrative frame is twice that of the household frame*, then the optimal allocation would have a sample size of 148,380 from the household frame and 955 from the administrative frame. Comparing the variances between the estimates of population proportion from the dual frame and the single-household frame, the relative variance R_V (dual-frame variance / household-frame variance) is,

$$R_V = 0.549.$$

This implies that the variance for the dual-frame design with these assumptions is 45 percent lower than the cost-equivalent variance for the household-frame design.

Looking at the same scenario again, when the cost of data collection using the administrative frame is ten (10) times that of the household frame, then the relative variance is:

$$R_V = 0.566,$$

showing very little difference from the above example.

TABLE 10-1 Effect of Population Proportion in Administrative Frame on Relative Variance

Proportion of Population with Attribute, P_A	Estimated Relative Variance, R_V
0.60	0.709
0.50	0.768
0.40	0.825
0.30	0.879
0.20	0.930

sampling and to estimate both the added costs and the relative efficiency of this proposed design. This research should also explore various practical strategies for integrating the samples chosen from the multiple frames. For instance, supplementary administrative lists could be developed within selected primary sampling units (PSUs; from the household frame) so that national administrative frames would not be necessary.

Lower-Cost Variations

Implementation of a new independent survey to measure rape and sexual assault would have a major impact on the current BJS budget. The panel is sensitive to these budget concerns, particularly in the current fiscal climate. Consequently, the panel considered several variations that would be less expensive than a full implementation of its recommendations for "best practices," while still making some improvements in the quality of the estimates. In doing so, we do not retreat from our charge to recommend a plan for best methods; rather, we present these lower-cost variations for BJS to consider within the realities of its budget priorities.

The panel first considered whether a traditional supplement would suffice in lieu of a completely independent survey.[3] The NCVS currently has a number of supplements, such as the 2008 Identity Theft Supplement. The major cost advantage to a supplement is that the survey is "piggy-backed" on the NCVS, using the same sample (or a subsample), the same field representatives, and the same survey structure. Individuals selected for a supplement would have been respondents on the NCVS survey over several waves of data collection. The demographics of the households and

[3] A survey supplement uses a supplemental questionnaire that is administered to a subset of respondents to the full survey, in this case the NCVS, in a separate data collection activity or wave. To avoid affecting the regular estimates, such supplements are usually administered to a subset of respondents who have rotated off the base survey.

individuals would be known. However, by the time they were asked to complete the supplement, their responses would still be in the context of a criminal victimization survey and they are unlikely to move beyond that context regardless of how the supplement is structured. Thus, if a supplement is used, it would need to be administered on the initial wave rather than on the households that are rotating off the NCVS.

A major problem with developing a supplement is that it would be based on the NCVS sample (or a subsample as described in "Enhanced Multiple-Frame Design" above). The design of that survey is very inefficient for measuring rape and sexual assault. With an independent survey, the sample design could be developed to be more efficient for measuring these specific criminal victimizations.

The NCVS is only one of several large household surveys conducted by the Census Bureau. The new survey may work better as a supplement to one of these other surveys rather than to the NCVS. For example, the American Community Survey (ACS) may be an appropriate vehicle. Because of the continuous design of the ACS, the Census Bureau could select individual household members (by appropriate demographic and geographic characteristics) who are at higher risk for rape and sexual assault. Using the ACS as a base for the new survey would avoid the context of a "crime."

Another lower-cost option would be to use the basic NCVS sample design (same PSUs and clusters) and select a large and separate sample for measuring rape and sexual assault. A household from this sample could be screened based on demographics of household members. Women ages 18-34 who have been shown to be of higher risk for rape and sexual assault would be oversampled as the respondents in this survey.

One way to reduce costs is to reduce the sample size or the frequency of data collection. BJS currently publishes annual estimates of the criminal victimizations of rape and sexual assault. The panel does not recommend less frequent data collection or publication, but one cost-saving variation would be to develop a formal rolling estimation process similar to that used for the ACS, which has a relatively modest annual sample size for the intent to produce small area estimates. This type of approach is currently used on an ad hoc basis on the NCVS for some special studies and research but not for annual estimates. The ACS uses a rolling estimation process in which multiple years of data (generally 3 years for most estimates) are pooled for estimation. Annually, data from the latest data collection are rolled into the estimates while data from the oldest year are rotated out. This process provides stable year-to-year estimates for small areas, even with modest sample sizes. Some more recent recommendations for the ACS are to use some time-series modeling across years rather than just a pooling of data. These ideas would also be useful to pursue with the estimates of rape and sexual assault.

Another variation would be for data from the NCVS to be used in conjunction with data from the recommended new survey to reduce overall data collection costs. The new survey could be viewed as a vehicle to measure the underestimation of rape and sexual assault on the NCVS. Data from both surveys could be modeled together to estimate this underestimation. The recommended new survey would be fielded on a fixed schedule, such as biennially or every 3 years. The NCVS could be modified to use behaviorally specific questions (as the panel recommends) and continue to be fielded annually. The NCVS estimates of rape and sexual assault would be adjusted for underestimation (based on the new survey) and continue to measure year-to-year change in criminal victimization rates. Considerable research would be needed to develop and calibrate these models.[4]

RECOMMENDATION 10-3 The Bureau of Justice Statistics should continue to publish annual estimates of rape and sexual assault criminal victimizations, using the recommended new survey to do so. However, if that is not possible, then the Bureau should conduct the recommended new survey on a fixed schedule, such as every 2 or 3 years, and use data from both the National Crime Victimization Survey and the new survey to calculate annual estimates of rape and sexual assault.

Other Recommended Design Characteristics

Longitudinal Component with Bounding

To measure the incidences of rape and sexual assault, the NCVS asks respondents to remember events in the past and to report only those that occurred within a specific reference period, currently 6 months. When a victim remembers a traumatic event such as a rape or sexual assault, he or she may remember the event as having occurred more recently than it actually did. (This could result in counting that event as part of a reference period when in reality it occurred before the beginning of that period, thus over-counting for that period.) This phenomenon is known as forward telescoping. A classic study examined the technique of "bounded recall"[5] to prevent the shifting of recalled events in time. In an experimental design, the study looked at collecting recall of expenditures on alterations and repairs made by resident homeowners. The study found that "unbounded recall of

[4]This would be a difficult task. The underreporting might not be uniform by subgroups, and BJS may lack sufficient data to accurately adjust by subgroup. Unless these adjustments could be extended beyond the estimates to the microdata, the differences would create difficulties in the analysis of rape and sexual assault victimizations.

[5]The previous interview is used to "bound" the reference period for recall.

expenditures for the preceding month involved substantial net forward telescoping of jobs into the recall period" (Neter and Waksberg, 1964, p. 43). The authors also concluded that shorter reference periods (1 month rather than 3 months) produced higher estimates of expenditures.

Because of the potential for telescoping, the NCVS originally did not use the data collected in its first wave in the estimation of criminal victimization. It currently uses data from initial interviews, with an adjustment to minimize overreporting during these initial contacts. (The adjustments are described in the "Estimation and Products" section in Chapter 4.)

In a recent analysis of NCVS data to look at the issues of telescoping, the role of the bounding interview, recency, and time-in-sample, Fay and Li (2010) found mixed results. More crimes were reported as having occurred in the month immediately preceding the interview than in other months in the reference period. The authors found that this "recency effect" is greater for violent crime than property crime. They concluded (Fay and Li, 2010, p. 1698):

> [T]he evidence presented here would encourage a re-examination of the issue of telescoping and the role of the bounding interview, . . . however, we recognize that BJS certainly has good reason to maintain the *status quo* for now, until other changes in the design are implemented in the future.

The panel strongly supports the need for further research. The appropriate reference period for recalling incidents of rape and sexual assault may be longer than for other criminal victimizations. Time-in-sample analysis may indicate the need for a survey with fewer waves.

Until more definitive work is done, the panel recommends the continuing use of bounded recall procedures. However, the panel has serious concerns about the current adjustments to first wave data to compensate for potential telescoping.

RECOMMENDATION 10-4 The recommended new survey should have a longitudinal structure with at least two waves to allow the use of bounded recall. Research should be conducted to determine an optimal length of reference period specifically for reporting rape and sexual assault victimizations. The Bureau of Justice Statistics should reassess the methodology used to adjust for forward telescoping if data from the bounding interview are used in estimation.

Neutral Context with Behaviorally Specific Wording

Concern about the context of crime in the NCVS and the use of terms such as "rape" and "sexual assault," and their potential effect to inhibit the reporting of incidents of rape and sexual assault, are discussed in-depth in Chapters 4 and 8. Some respondents may not view their victimization as criminal. Or they may have decided not to report the incident to police as a crime and now have concerns about reporting it on a government crime survey. When asked specifically about "rape" and "sexual assault," survey respondents may not consistently or accurately understand those terms. Research has shown that a change to behaviorally specific questions increases reporting of the criminal victimizations (Fisher, 2009). As detailed in Chapter 8, the context of a crime survey is likely to inhibit positive responses (Conclusion 8-7 in Chapter 8), and the use of behaviorally specific questions would likely lead to more accurate responses (Conclusion 8-6 in Chapter 8).

RECOMMENDATION 10-5 The questionnaire and protocols for the recommended new survey should have a neutral context, such as a health survey. The Bureau of Justice Statistics should explore several neutral alternatives while continuing to use both a victimization screening questionnaire and an incident report. The questions on both of these instruments should be reworded to incorporate behaviorally specific questions.

Self-Administered Mode of Data Collection with One Household Member Selected

It is critically important to provide privacy to respondents when asking them to recall the details of a rape or sexual assault. The current NCVS does not provide this privacy, and this fact is one of the primary reasons for the panel's recommendation to conduct a separate survey to measure incidence rates. The lack of privacy in the current NCVS is rooted in the fact that everyone in the household is interviewed and therefore knows the questions that are being asked, and there is an oral interview taking place that might be overheard. A second issue in dealing with these sensitive questions is that respondents may feel that reporting their criminal victimization to an interviewer may be socially undesirable, so they do not report accurately. (These issues are discussed in-depth in Chapter 8.)

These characteristics of the NCVS lead to the panel's recommendation for a self-administered mode of collecting the information and a single-respondent design. There are significant cost considerations in moving to a survey with a single respondent per household. It takes considerable resources to make the initial contact with a household, and the current NCVS then has multiple respondents as a result of that contact. Thus, the sampling error will increase under a single-respondent design unless a larger sample of households is selected or multiple-frame sampling provides efficiency in finding individuals at risk for this victimization (as described in "Enhanced Multiple-Frame Design" above). The panel is confident that having a single-respondent design would be less problematic in the single-purpose survey for rape and sexual assault than it would be for the entire NCVS, which is obtaining information on multiple types of victimizations from different family members. This is a topic for which additional research would be helpful.

The selection of a single respondent within a household should not be made with equal probabilities of selection. Instead, individuals whose demographics would put them at greater risk for sexual criminalization (females, certain age groups, etc.) would have higher probabilities of selection. This would be straightforward in a survey specifically designed for measuring rape and sexual assault. It would be very difficult to do for an omnibus survey because the demographics more "at risk" would be different for different types of victimizations.

Advances in technology have eliminated many of the past concerns about self-administered surveys. In particular, audio computer-assisted self-administered interviewing (ACASI) has evolved with the proliferation of computers and computer-assisted survey techniques. In an ACASI survey, the interviewer obtains the respondent's consent and cooperation and provides basic instruction. The respondent is then seated in front of a laptop or tablet computer and puts on a set of earphones. The survey questions appear visually on the computer screen and orally through the earphones simultaneously. The respondent can control the speed of each input source and ask additional questions of the interviewer as needed.

The use of ACASI technology will increase privacy. However, the panel is very concerned that the use of this technology alone will not provide the privacy needed in households in which the victim is concerned about other household members knowing that he or she is answering questions about rape and sexual assault. Thus the panel continues to support the single-respondent design even with the use of ACASI technology.

The Internet offers another approach to self-administered surveys, but this mode lacks the presence of an interviewer to encourage response and answer questions. An Internet-based survey may be an acceptable variation for certain respondents on subsequent waves following an initial in-person contact.

RECOMMENDATION 10-6 The recommended new survey should be conducted in a self-administered mode. The wave 1 contact would involve a personal visit and audio computer-assisted self-administered interviewing technology. Only one individual in a selected household should be selected for this survey to increase the respondent's privacy.

Modified Definition of Rape

BJS currently uses a fairly broad definition of rape and sexual assault (see Box 8-1 in Chapter 8). However, the definitions do not include the incapacity to consent to sexual activities. Under most current laws, the capacity to consent is affected by the age of the victim, mental capability of the victim, and intoxication (see Chapter 2). This component should be included in the basic definitions: examples of potential revised definitions are shown in Box 10-2. As with any major definitional change, this change is likely to affect the estimates. These effects should be investigated, and a bridge with the older definition should be used as the new definitions are phased in.

RECOMMENDATION 10-7 The Bureau of Justice Statistics' definition of rape and sexual assault should be expanded to include victimizations when the victim does not have the capacity to consent to the sexual actions of the offender.

BJS publishes its definitions and questionnaires on its website. However, the link between the answers given to various questions and the overall classification of the incident as a crime is not clear. Such a link would allow a more transparent understanding of the classification of incidents reported on the survey. The flowchart presented in Figure 7-3 (in Chapter 7) may be useful as a starting point for this link.

BOX 10-2
Proposed Revisions to BJS Definitions of Rape and Sexual Assault

Rape—Forced sexual intercourse including ~~both~~ psychological coercion, ~~as well as~~ physical force, ***and the victim's inability to consent***. Forced sexual intercourse means vaginal, anal or oral penetration by the offender(s). This category also includes incidents where the penetration is from a foreign object such as a bottle. Includes attempted rapes, male as well as female victims, and both heterosexual and homosexual rape. Attempted rape includes verbal threats of rape.

Sexual Assault—A wide range of criminal victimizations, separate from rape or attempted rape. These crimes include attacks or attempted attacks generally involving unwanted sexual contact between victim and offender. Sexual assaults may or may not involve force and include such things as grabbing or fondling. Sexual assault also includes verbal threats ***and situations where the victim does not have the capability to consent***.

NOTES: Strikethroughs are deletions. Text in boldface italics are additions.
SOURCE: Panel modification to current NCVS definitions (Bureau of Justice Statistics, n.d.-b).

RECOMMENDATION 10-8 The Bureau of Justice Statistics should make more transparent the link between responses on the survey screener and incident reports and the final type of crime classification of those incidents of potential rape and sexual assault.

Survey Covariates

The NCVS and the recommended new independent survey can provide data that increase understanding of the frequency and context under which criminal victimizations occur. This improved understanding often comes from the research performed by criminologists, sociologists, and others using data from the NCVS. Having demographic variables and other relevant covariates available in the datasets enhances the value of the criminal victimization data.

The design of the recommended new survey with a focus on sexual violence provides an opportunity to review and enhance the set of covariates that are collected along with the criminal victimization variables. Researchers who study criminal victimizations have identified certain types of information about people's lifestyle and routine activities that can provide insight into, and perhaps lead to identifying causal relationships between, the context and likelihood of future victimization (Fisher et al., 1998; Schreck, Stewart, and Fisher, 2006). It will be important to involve data users to assist in identifying a useful set of covariates.

RECOMMENDATION 10-9 The recommended new survey should include a number of covariates to add to the richness of the data set for data analysis. The Bureau of Justice Statistics should hold an expert-user workshop as it develops the new survey. A major purpose of the workshop would be to obtain advice on the covariates that could best improve the usefulness of the dataset for research, advocacy, and policy purposes.

Alternative Procedures for Series Victimizations

Series victimization is defined on the NCVS as the situation when a single respondent reports six or more separate but similar victimizations over the reference period but is unable to recall these events individually or describe them in detail to the interviewer. In 2011, BJS implemented a procedural change in how these series victimizations are handled on the NCVS. Chapter 7 discusses the impact of this change on the estimation of rape and sexual assault. The impact of the resulting outliers was sizable, resulting in large increases in the estimates of rape and sexual assault and the standard errors, and much greater year-to-year fluctuation. With a new survey that focuses on measuring rape and sexual assault, there is an opportunity to reexamine the procedures that should be used in this survey.

The panel sees a key first step is to better understand the subpopulation that is experiencing multiple rapes and sexual assaults within a short period of time. This understanding would include the demographic and lifestyle characteristics of these victims as well as the size of the subpopulation. A second step is to investigate whether series victimizations are reported with sufficient quality to use in the estimation process, given the concerns about response error and other measurement errors discussed in Chapter 8. If not, alternative data collection procedures should be developed based on research (including cognitive testing) designed to better understand how to retrieve information on series victimizations. This research could lead to both better data collection procedures as well as data adjustment techniques to improve the quality of data that are collected.

Once the data are determined to be of sufficient quality, thought should be given to what is the primary goal of the survey—estimation within a single year, estimation of totals or rates across several years, or estimation of the annual change in rates or totals. If the latter two goals are of primary interest, then consideration might be given to spreading outlier respondents' estimation weights across several years. This would have an effect akin to using a running average for the outliers each year.

Here is an example of a possible procedure for smoothing the responses over 3 years. Let W denote the respondent's survey weight, and let R denote the number of rape and sexual assault victimizations reported by the respondent. A cap, C, on R may be imposed, so that R is truncated to the cap. BJS currently uses $C = 10$. Let the capped value of R be denoted by $S = \min(R, C)$. Let D denote a threshold value. If the product WS (weight times [capped] data value) exceeds D, then the respondent is included in the current wave with a modified weight such as

$$W' = (1/3)W + (2/3)(D/S)$$

A separate record is created for the respondent in the survey data for the wave 12 months later and for the wave 24 months later with the same data values and modified weight

$$W'' = (1/3)W - (1/3)(D/S).$$

If the respondent is included in the sample in one of the latter waves, then the respondent would have more than one survey record for the wave. How to choose the number of years, the cap, and the threshold value are questions that would need to be researched. The procedure as described is fairly simple and could be refined.

The panel recommends continuing research into improved methods for accurately accounting for series victimizations while minimizing the year-to-year effect caused by the current outlier adjustment procedure.

Recommendation 10-10 The Bureau of Justice Statistics should reassess the methodological change made to the National Crime Victimization Survey regarding series victimization and should investigate alternative procedures that are more effective in respect to measurements of rape and sexual assault. This reassessment should involve formal input by experts on outlier adjustment techniques and by data users who can help assess the relative tradeoffs in quality.

SPECIALIZED TRAINING AND MONITORING

There is a clear need for improved training for Census Bureau field representatives who administer the NCVS (see Chapter 8). Training needs to be a high-priority activity even in times of tight budgets. The development of a separate independent survey will provide an opportunity to have focused, specialized training for the field representatives working on this survey. Their role will be different with the use of ACASI technology. They will need to have training on how to introduce this survey and encourage response. They will need special training regarding the topics of rape and sexual assault, the words to use in answering questions, and the correct level of emotional support to show.

The panel also recommends that the interaction between the field representative and the respondent be recorded using CARI (computer-assisted recorded interviewing) technology. These recordings can be used effectively to monitor the interaction process and to identify areas where retraining (of individual interviewers or more generally) will be important. It is possible that the use of a recording device may affect survey response. Although existing research does not indicate that CARI reduces response rate (Biemer et al., 2000; Thissen et al., 2009), it would be prudent to consider this an open research question to be investigated as the survey design is developed.

Recommendation 10-11 The Census Bureau and the Bureau of Justice Statistics should provide specialized training to field representatives on how to assist the respondents and answer questions on the sensitive subjects of rape and sexual assault. The interaction between respondents and field representatives should be recorded using computer-assisted recorded interviewing technology.

RESEARCH

Many of the recommendations presented in this report will need additional research and development before implementation.

Recommendation 10-12 The Bureau of Justice Statistics should conduct a series of coordinated research investigations to enable it to resolve each of the following specific issues in developing the design for the recommended stand-alone survey on rape and sexual assault.

The panel recommends that BJS develop an ongoing program of research addressing the following 11 topics.

1. The cost-efficiency of introducing disproportionate stratified sampling of those at higher risk for rape and sexual assault victimization. This includes determining (i) which variables to use for stratification at each stage of selection and (ii) the possible utility of disproportionate sample allocation, particularly in sampling residences from nonoverlapping residential listings.
2. The cost-efficiency of supplementing the standard area household sampling frame with one or more frames derived from administrative sources. This includes determining (i) whether it would be more cost-effective to select the sample from a single NCVS-like household sampling frame alone, or from a household sampling frame plus one or more overlapping frames from administrative sources of possible victims and (ii) regardless of the feasibility of the multiple-frame option, whether design integration with a household survey conducted by another federal statistical agency (e.g., the American Community Survey conducted by the Census Bureau) would be a statistically and practically feasible way to generate the household sample that will be needed for the recommended stand-alone survey.
3. The best estimation approach to deal with telescoping effects that arise from the use of bounded questions in a longitudinal setting. This research should include the examination of the potential to use a longer reference period (possibly 1 year) for the new survey.
4. The overall effect of changing the "context" of the survey from that of crime to a more neutral context.
5. The effects of following a neutral/behavioral orientation for questions used to screen for rape and sexual assault victimization. This research should also include the examination of potentially useful covariates for predicting rape and sexual assault criminal victimization.

6. The joint sampling and measurement error implications of self-administration of a single respondent chosen in each participating household. This research should also include the exploration of the use of the Web for self-administration after the first wave.
7. The effect of expanding the definition of rape and sexual assault to include those without the capacity to give their consent to the offender.
8. The error and cost implications of improved training and supervision of field representatives.
9. Ways to improve estimation in the presence of series victimization. This includes an examination of the characteristics of the subpopulation that is sexually victimized on a frequent basis (series victimization) to provide data to better represent this subpopulation in the overall victimization rates.
10. Determine whether effective models can be built to estimate the underreporting on the NCVS, using data from the new survey that is fielded on a periodic basis.
11. Conduct further research on issues related to collecting data on rape and sexual assault criminal victimization from adolescents (12-17 years of age) because of their relatively high risk of criminal victimization.

COMMUNICATION WITH DATA USERS

Principles and Practices for a Federal Statistical Agency identifies as one of its key principles the need to maintain credibility among data users. It states the following (National Research Council, 2013b, p. 13):

> Credibility derives from the respect and trust of users in the statistical agency and its data. Such respect results not only from an agency's production of data that merit acceptance as relevant, accurate, timely, and free from political and other undue external influence, but also from many aspects of an agency's policies and practices. Key among these are wide dissemination of data on an equal basis to all users; openness about the sources and processes used to produce data and the limitations of the data; commitment to quality and professional practice; a strong internal and external evaluation program to assess and improve an agency's data systems; a willingness to understand and strive to meet user needs, even though users may not clearly articulate their needs; and a posture of respect and trust in the users of an agency's data.

In pursuing this general principle, *Principles and Practices* recommends that these agencies engage in these specific practices (National Research Council, 2013b, pp. 19-20):

- seek advice on data concepts, content, processing, estimation, products, and documentation from a wide spectrum of data users, as well as from professional and technical subject-matter and methodological experts, using a variety of formal and informal means of communication that are appropriate to the types of input sought;
- seek advice on its statistical programs and priorities from external groups, including those with relevant subject-matter and technical expertise;
- keep abreast of and use modern statistical theory and sound statistical practice in all technical work; and
- document concepts, definitions, data collection methodologies, and measures of uncertainty and discuss sources of error in reports and other data releases to the public.

In its review of U.S. justice statistics (*Ensuring the Quality, Credibility, and Relevance of U.S. Justice Statistics*) the National Research Council (2009, pp. 250-251) focused on these principles and practices and made several specific recommendations regarding the need for BJS to maintain both formal and informal communication with its data users.

- BJS should regularly convene ad hoc stakeholder workshops to suggest areas of immediate data needs (Recommendation 5.7), and
- BJS should establish an Advisory Group under the Federal Advisory Committee Act to provide guidance to BJS on the addition of new data collection efforts and the modification of current ones in light of needs identified by the group (Recommendation 5.8).

BJS has reached out to data users in a number of useful ways. Over the years, BJS has held expert workshops that were well received by data users. Unfortunately, the frequency of the workshops has decreased in recent years because of budget constraints. BJS has made use of expertise and advice from the American Statistical Association's Committee on Law and Justice Statistics. BJS has also sponsored an annual 4-week summer workshop at the University of Michigan's Inter-university Consortium for Political and Social Research to train data users on analyses techniques for their datasets. The panel commends BJS for those efforts and recommends more.

This panel endorses the recommendation from the previous NRC report discussed above and repeats it as a recommendation.

Recommendation 10-13 The Bureau of Justice Statistics should establish a permanent advisory committee under the Federal Advisory Committee Act to provide guidance on user issues, ongoing program and research priorities, and the implementation of new methodological advances. This committee should advise on all of the survey programs of

the Bureau of Justice Statistics and not be limited to the National Crime Victimization Survey or the measurement of rape and sexual assault.

Going beyond the formal advisory committee for its entire program, BJS should also convene a regular data user conference focused on the measurement of criminal victimizations, including the measurement of rape and sexual assault.

Recommendation 10-14 The Bureau of Justice Statistics should hold a regular (annual or biennial) data users' conference for users of the Bureau of Justice Statistics criminal victimization data. The conference would have four objectives: (1) to provide an update on new or planned program changes, (2) to facilitate informal communication between data users and the Bureau of Justice Statistics staff on issues important to those data users, (3) to provide training on the use of publicly available microdata, and (4) to provide a vehicle for data users to showcase academic papers using these data. This data users' conference might be held in conjunction with a professional association meeting.

Going beyond a regular data users' conference, BJS should more closely follow the practice identified in *Principles and Practices* when contemplating a major methodological change in one of its major surveys (National Research Council, 2013b, p. 9, emphasis added):

> [S]eek advice on data concepts, content, processing, estimation, products, and documentation from a wide spectrum of data users, as well as from *professional and technical subject-matter and methodological experts*, using a variety of formal and informal means of communication that are appropriate to the types of input sought.

Specifically, the panel recommends that BJS take the following steps before making a major change:

- develop the conceptual background for change, along with alternative sampling, estimation, and/or survey methodology changes that are being considered;
- formally review the issues with data users to decide on a strategic direction for the change—this could be done during a data users' conference;
- formally review the proposed sampling, estimation, and/or survey methodology changes with the broader statistical community to obtain input on best methodologies for implementing the proposed changes;

- formally test and evaluate the selected new procedure(s); and
- clearly communicate with data users at each step.

Recommendation 10-15 The Bureau of Justice Statistics should follow five steps when contemplating a major methodological change in one of its major surveys: (1) develop the conceptual background for the need for change and alternative sampling, estimation, or survey methodologies; (2) formally review these concepts with a broad set of data users to decide on a strategic direction; (3) formally review the statistical and survey methodological issues and proposed changes with technical experts in the broader statistical community; (4) formally test and evaluate the new procedures, their feasibility, and their impact; and (5) clearly communicate with data users at each step.

The panel offers two concrete examples regarding the need for the above recommendation: the recent change to account for series victimization and an earlier decision to use data from the bounding interview in estimation. In both cases, BJS concluded that a methodology change was needed and had informal, if not formal, discussions with a number of data users who weighed in on the issues. However, the third step is also important, and it appears to have received less attention. In the example of series victimization, BJS appears not to have engaged fully with specialists on outliers in the statistical community to help evaluate a wider range of alternative procedures to adjust (or not adjust) for these outliers. In the example of the bounding interview, the current adjustments that involve weighting the first wave data down to the average level of waves 2-7 are suspect. As more recent analysis has shown, the "recency" effect, along with notable attrition rates over the life of the survey, raises the question of whether these adjustments to the data are the best approach. Again, engaging more statistical expertise to examine alternatives would have been appropriate when this adjustment procedure was considered.

Following the decision to change, communication with data users needs to be frequent and clear. For example the publication *Criminal Victimizations, 2011* (Bureau of Justice Statistics, 2012a) stated that the series victimization change was being implemented in that, and future, publications. The panel did not find a notice that the change was being made retroactively on the online database. More and frequent communication is always a good policy.

WRAP-UP

BJS has a very important mission to provide estimates of criminal victimizations within the United States, both annual rates of those victimizations and the change between years. It employs the NCVS, a large omnibus victimization survey, as a critical tool for accomplishing that mission. Addressing data user questions about its current methodology for measuring the incidence of rape and sexual assault, the BJS sought advice from the National Research Council, which led to the creation of this panel. The panel concludes that "best practices" for measuring rape and sexual assault on BJS household surveys would involve a decoupling of that measurement from the NCVS, and it provided guidelines for making that change. The panel applauds BJS for its openness in addressing these issues and is confident that it can use the analyses and recommendations in this report to move toward continuous quality improvements in measuring these injurious victimizations.

References

AEquitas. (2012). *Rape and Sexual Assault Analysis and Laws.* Available: http://www.aequitasresource.org/Rape_and_Sexual_Assault_Analyses_and_Laws.pdf [October 2013].

American Association for Public Opinion Research. (n.d.). *Response Rate: An Overview.* Available: http://www.aapor.org/Response_Rates_An_Overview1.htm#.Ul6jkWRmtHg [October 2013].

Amir, M. (1971). *Patterns in Forcible Rape.* Chicago, IL: University of Chicago Press.

Aquilino, W., Wright, D., and Supple, A. (2000). Response effects due to bystander presence in CASI and paper-and-pencil surveys of drug use and alcohol use. *Substance Use and Misuse,* 35, 845-867.

Bachman, R. (1993). Predicting the reporting of rape victimizations: Have reforms made a difference? *Criminal Justice and Behavior, 20,* 254-270.

Bachman, R. (1998). The factors related to rape reporting behavior and arrest: New evidence from the National Crime Victimization Survey. *Criminal Justice and Behavior, 25*(1), 8-25.

Bachman, R. (2012). *Measuring Rape and Sexual Assault: Successive Approximations to Consensus.* Paper commissioned by the National Research Council, Panel on Measuring Rape and Sexual Assault in the Bureau of Justice Statistics Household Surveys. Available: http://sites.nationalacademies.org/DBASSE/CNSTAT/CurrentProjects/DBASSE_073316#.UVR8VltmtJE [March 2013].

Bachman, R., and Taylor, B. (1994). The measurement of family violence and rape by the redesigned National Crime Victimization Survey. *Justice Quarterly, 11*(3), 499-512.

Barnett, V., and Lewis, T. (1994). *Outliers in Statistical Data* (third ed.). Chichester, England: John Wiley & Sons.

Barnett-Ryan, C. (2007). Introduction to the Uniform Crime Reporting Program. In J. Lynch and L. Addington (Eds.), *Understanding Crime Statistics: Revisiting the Divergence of the NCVS and UCR* (pp. 55-92). Cambridge, NY: Cambridge University Press.

Baumer, E.P., and Lauritsen, J.L. (2010). Reporting crime to the police 1973-2005: A multivariate analysis of long-term trends in the National Crime Victimization Survey. *Criminology, 48*(1), 131-185.

Belknap, J. (2001). *The Invisible Woman: Gender, Crime, and Justice* (second ed.). Toronto, Canada: Wadsworth.

Berkowitz, S. (2011). *NCVS Data: Its Uses and Shortcomings.* Presentation at the meeting of the Panel on Measuring Rape and Sexual Assault in the Bureau of Justice Statistics Household Surveys, National Academy of Sciences, Washington, DC.

Biderman, A. (1981). Sources of data for victimology. *The Journal of Criminal Law and Criminology, 72,* 789-817.

Biderman, A., and Reiss, A. (1967). On exploring the "dark figure" of crime. *The ANNALS of the American Academy of Political and Social Science, 374*(1), 1-15.

Biemer, P. (2010). Overview of design issues: Total survey error. In P. Marsden and J. Wright (Eds.), *Handbook of Survey Research* (second ed., pp. 27-58). Bingley, UK: Emerald Group.

Biemer, P., Hergert, D., Morton, J., and Willis, W. (2000). *The Feasibility of Monitoring Field Interview Performance Using Computer Audio Recorded Interviewing (CARI).* Proceedings of the Survey Research Methods Section, American Statistical Association. Available: http://www.amstat.org/sections/srms/proceedings/papers/2000_183.pdf [October 2013].

Black, M.C., Basile, K.C., Breiding, M.J., Smith, S.G., Walters, M.L., Merrick, M.T., Chen, J., and Stevens, M.R. (2011). *The National Intimate Partner and Sexual Violence Survey (NISVS): 2010 Summary Report.* Atlanta, GA: National Center for Injury Prevention and Control, Centers for Disease Control and Prevention.

Blumberg, S., and Luke, J. (2012). *Wireless Substitution: Early Release of Estimates from the National Health Interview Survey, January-June 2012.* Available: http://www.cdc.gov/nchs/data/nhis/earlyrelease/wireless201212.pdf [October 2013].

Bradburn, N. (1983). Response effects. In P. Rossi, J. Wright, and A. Anderson (Eds.), *Handbook of Survey Research* (pp. 289-328). New York: Academic Press.

Bradburn, N., Sudman, S., and Blair, E. (1979). *Improving Interview Method and Questionnaire Design.* San Francisco, CA: Jossey-Bass.

Brooks, C., and Bailar, B. (1978). *An Error Profile: Employment as Measured by the Current Population Survey.* Statistical Working Paper 3. Washington, DC: U.S. Office of Management and Budget.

Brunson, R., and Miller, J. (2006). Gender, race, and urban policing: The experiences of African American youth. *Gender & Society, 20*(4), 531-552.

Bureau of Justice Statistics. (1989). *Redesign of the National Crime Survey.* Available: http://www.bjs.gov/index.cfm?ty=pbdetail&iid=3771 [August 2013].

Bureau of Justice Statistics. (1991). *Criminal Victimization in the United States, 1990.* Available: http://www.bjs.gov/content/pub/pdf/cvus90.pdf [October 2013].

Bureau of Justice Statistics. (1992). *Criminal Victimization, 1991.* Available: http://www.bjs.gov/content/pub/pdf/cv91.pdf [October 2013].

Bureau of Justice Statistics. (1993). *Criminal Victimization, 1992.* Available: http://www.bjs.gov/content/pub/pdf/Cv92.pdf [October 2013].

Bureau of Justice Statistics. (1994). *Criminal Victimization, 1993.* Available: http://www.bjs.gov/content/pub/pdf/Cv93.pdf [October 2013].

Bureau of Justice Statistics. (1995). *Criminal Victimization, 1994.* Available: http://www.bjs.gov/content/pub/pdf/Cv94.pdf [October 2013].

Bureau of Justice Statistics. (1996). *Criminal Victimization in the United States, 1995.* Available: http://www.bjs.gov/content/pub/pdf/cvus95.pdf [October 2013].

Bureau of Justice Statistics. (1997). *Criminal Victimization, 1996.* Available: http://www.bjs.gov/content/pub/pdf/cv96.pdf [October 2013].

Bureau of Justice Statistics. (1998). *Criminal Victimization, 1997.* Available: http://www.bjs.gov/content/pub/pdf/cv97.pdf [October 2013].

Bureau of Justice Statistics. (1999). *Criminal Victimization, 1998.* Available: http://www.bjs.gov/content/pub/pdf/cv98.pdf [October 2013].

Bureau of Justice Statistics. (2000). *Criminal Victimization, 1999.* Available: http://www.bjs.gov/content/pub/pdf/cv99.pdf [October 2013].

Bureau of Justice Statistics. (2001). *Criminal Victimization, 2000.* Available: http://www.bjs.gov/content/pub/pdf/cv00.pdf [October 2013].

Bureau of Justice Statistics. (2002a). *Criminal Victimization, 2001.* Available: http://www.bjs.gov/content/pub/pdf/cv01.pdf [October 2013].

Bureau of Justice Statistics. (2002b). *Rape and Sexual Assault: Reporting to Police and Medical Attention, 1992-2000.* Available: http://www.bjs.gov/content/pub/pdf/rsarp00.pdf [October 2013].

Bureau of Justice Statistics. (2003). *Criminal Victimization, 2002.* Available: http://www.bjs.gov/content/pub/pdf/cv02.pdf [October 2013].

Bureau of Justice Statistics. (2004). *Criminal Victimization, 2003.* Available: http://www.bjs.gov/content/pub/pdf/cv03.pdf [October 2013].

Bureau of Justice Statistics. (2005). *Criminal Victimization, 2004.* Available: http://www.bjs.gov/content/pub/pdf/cv04.pdf [October 2013].

Bureau of Justice Statistics. (2006). *Criminal Victimization, 2005.* Available: http://www.bjs.gov/content/pub/pdf/cv05.pdf [October 2013].

Bureau of Justice Statistics. (2007). *Criminal Victimization, 2006.* Available: http://www.bjs.gov/content/pub/pdf/cv06.pdf [October 2013].

Bureau of Justice Statistics. (2008a). *Criminal Victimization, 2007.* Available: http://www.bjs.gov/content/pub/pdf/cv07.pdf [October 2013].

Bureau of Justice Statistics. (2008b). *Survey Methodology for Criminal Victimization in the United States.* Available: http://bjs.ojp.usdoj.gov/content/pub/pdf/ncvs_methodology.pdf [March 2013].

Bureau of Justice Statistics. (2009). *Criminal Victimization, 2008.* Available: http://www.bjs.gov/content/pub/pdf/cv08.pdf [October 2013].

Bureau of Justice Statistics. (2010). *Criminal Victimization, 2009.* Available: http://www.bjs.gov/content/pub/pdf/cv09.pdf [October 2013].

Bureau of Justice Statistics. (2011). *Criminal Victimization, 2010.* Available: http://www.bjs.gov/content/pub/pdf/cv10.pdf [October 2013].

Bureau of Justice Statistics. (2012a). *Criminal Victimization (2011).* Available: http://www.bjs.gov/content/pub/pdf/cv11.pdf. [October 2013].

Bureau of Justice Statistics. (2012b). *Victimizations Not Reported to the Police, 2006-2010.* Available: http://bjs.gov/content/pub/pdf/vnrp0610.pdf [March 2013].

Bureau of Justice Statistics. (n.d.-a). *NCVS Victimization Analysis Tool.* Available: http://www.bjs.gov/index.cfm?ty=nvat [March 2013].

Bureau of Justice Statistics. (n.d.-b). *Terms and Definitions–Victims.* Available: http://bjs.ojp.usdoj.gov/index.cfm?ty=tdtp&tid=9 [March 2013].

Bureau of Justice Statistics. (n.d.-c). *Mission.* Available: http://bjs.ojp.usdoj.gov/index.cfm?ty=abu [March 2013].

Bureau of Justice Statistics. (n.d.-d). *NCVS Basic Screening Questionnaire 2009-2011.* Available: http://www.bjs.gov/content/pub/pdf/ncvs109.pdf [July 2013].

Cannell, C.F., Miller, P.V., and Oksenberg, L. (1981). Research on interviewing techniques. In S. Leinhardt (Ed.), *Sociological Methodology* (pp. 389-437). San Francisco, CA: Jossey-Bass.

Cantor, D., and Lynch, J. (2000). Self-report surveys as measures of crime and criminal victimization. In R. Kaminski and N. La Vigne (Eds.), *Criminal Justice 2000* (vol. 4, pp. 85-138). Washington, DC: U.S. Department of Justice, National Institute of Justice.

Carbone-Lopez, K., Slocum, L., and Kruttschnitt, C. (Forthcoming). Police wouldn't give you no help: Female offenders on reporting sexual assault to police. Submitted to *Violence Against Women.*

Catania, J. (1997). A model for investigating respondent-interviewer relationships in sexual surveys. In J. Bancroft (Ed.), *Researching Sexual Behavior: Methodological Issues* (pp. 417-443, The Kinsey Institute Series, vol. 5). Bloomington: Indiana University Press.

Chakrabarty, R., and Torres, G. (1996). *American Housing Survey: A Quality Profile.* (#H121/95-1.) Washington, DC: U.S. Department of Housing and Urban Development, U.S. Census Bureau. Available: http://www.census.gov/prod/2/constr/h12195-1.pdf [June 2013].

Chromy, J., and Wilson, D. (2012). *Multiple Frame Approaches to Identify and Survey Victims of Rape and Sexual Assault.* Paper commissioned by the National Research Council Panel on Measuring Rape and Sexual Assault in the Bureau of Justice Statistics Household Surveys, Washington, DC.

Clay-Warner, J., and Burt, C.H. (2005). Rape reporting after reforms: Have times really changed? *Violence Against Women, 11*(2), 150-176.

Cochran, W.G. (1977). *Sampling Techniques* (third ed.). New York: John Wiley & Sons.

Dempsey, M. (2010, September 14). *Rape in the United States: The Chronic Failure to Report and Investigate Rape Cases.* Hearing before the Senate Committee on the Judiciary, Subcommittee on Crime and Drugs. Available: http://www.judiciary.senate.gov/pdf/10-09-14MaddenDempseyTestimony.pdf [March 2013].

Diamond, L., and Emerson, B. (2012, November 18). *Sexual Assaults on College Campuses: A Problem Nationwide.* Available: http://www.ajc.com/news/news/sexual-assaults-on-college-campuses-a-problem-nati/nS8TT [March 2013].

Doyle, P., and Clark, C. (2001, May). *Quality Profiles and Data Users.* Paper presented at the International Conference on Quality in Official Statistics, Stockholm, Sweden.

Dykema, J., Diloreto, K., Price, J., White, E., and Schaeffer, N. (2012). ACASI gender-of-interviewer voice effects on reports to questions about sensitive behaviors among young adults. *Public Opinion Quarterly, 76*(2), 311-325.

Estrich, S. (1987). *Real Rape.* Cambridge, MA: Harvard University Press.

Fay, R., and Li, J. (2010). *The Effects of Unbounded Interviews, Time in Sample, and Recency on Reported Crime in the National Crime Victimization Survey.* Proceedings of the Survey Research Methods Section, American Statistical Association. Available: http://www.amstat.org/sections/srms/proceedings/y2010/Files/306795_57200.pdf [January 2014].

Federal Bureau of Investigation. (1997). *Crime in the United States, 1997.* Available: http://http://www.fbi.gov/about-us/cjis/ucr/crime-in-the-u.s/1997 [October 2013].

Federal Bureau of Investigation. (2004). *Uniform Crime Reporting Handbook.* Available: http://www.fbi.gov/about-us/cjis/ucr/additional-ucr-publications/ucr_handbook.pdf/view [March 2013].

Federal Bureau of Investigation. (2007). *Crime in the United States, 2007.* Available: http://www.fbi.gov/about-us/cjis/ucr/crime-in-the-u.s/2007/crime-in-the-u.s.-2007 [March 2013].

Federal Bureau of Investigation. (2011). *Crime in the United States, 2011.* Available: http://www.fbi.gov/about-us/cjis/ucr/crime-in-the-u.s/2011/crime-in-the-u.s.-2011 [March 2013].

Federal Bureau of Investigation. (2012, January 6). *Attorney General Eric Holder Announces Revisions to the Uniform Crime Report's Definition of Rape* [Press release]. Available: http://www.fbi.gov/news/pressrel/press-releases/attorney-general-eric-holder-announces-revisions-to-the-uniform-crime-reports-definition-of-rape [March 2013].

Federal Bureau of Investigation. (2013). *Supporting Statement to the Office of Management and Budget, 1110-001 Return A-Monthly Return of Offenses Known to the Police.* Available: http://www.reginfo.gov/public/do/PRAViewDocument?ref_nbr=200910-1110-001 [March 2013].

Federal Bureau of Investigation. (n.d.-a). *A Word about UCR Data*. Available: http://www.fbi.gov/about-us/cjis/ucr/word [March 2013].

Federal Bureau of Investigation. (n.d.-b). *Uniform Crime Reports*. Available: http://www.fbi.gov/about-us/cjis/ucr/ucr [March 2013].

Federal Bureau of Investigation. (n.d.-c). *UCR Data Tool*. Available: http://www.fbi.gov/about-us/cjis/ucr/ucr [March 2013].

Felson, R.B., and Pare, P.P. (2005). The reporting of domestic and sexual assault by nonstrangers to police. *Journal of Marriage and the Family, 67*(3), 597-610.

Fenton, J. (2010, June 27). *City Rape Statistics, Investigations Draw Concern*. Available: http://articles.baltimoresun.com/2010-06-27/news/bs-md-ci-rapes-20100519_1_fbi-data-mayor-orders-review-detectives [March 2013].

Fisher, B. (2009). The effects of survey question wording on rape estimates: Evidence from a quasi-experimental design. *Violence Against Women, 15*(2), 133-147.

Fisher, B., and Cullen, F. (1999). *Violent Victimization Against College Women: Result from a National-Level Study.* Unpublished report. Washington, DC: U.S. Department of Justice, Bureau of Justice Statistics.

Fisher, B., Sloan, J., Cullen, F., and Lu, C. (1998). Crime in the ivory tower: The level and sources of student victimization. *Criminology, 36*(3), 671-710.

Fisher, B., Cullen, F., and Turner, M. (2000). *The Sexual Victimization of College Women*. (NCJ #182369.) Washington, DC: U.S. Department of Justice, National Institute of Justice. Available: https://www.ncjrs.gov/pdffiles1/nij/182369.pdf [March 2013].

Fisher, B., Daigle, L., Cullen, F., and Turner, M. (2003). Reporting sexual victimization to the police and others: Results from a national-level study of college women. *Criminal Justice and Behavior, 30*(1), 6-38.

Gibson, C., Shapiro, G., Murphy, L., and Stanko, G. (1978). *Interaction of Survey Questions as It Relates to Interviewer-Respondent Bias*. Proceedings of the Survey Research Methods Section, American Statistical Association. Available: http://www.amstat.org/sections/srms/Proceedings [January 2014].

Greenberg, M., and Ruback, B. (1992). *After the Crime: Victim Decision Making*. New York: Plenum Press.

Hall, J., Carlson, B., and CyBulski, K. (2011). *RDD Unplugged: Findings from a Household Survey Using a Cell Overlap Design*. Working paper. Available: http://www.mathematica-mpr.com/publications/pdfs/RDD_unplugged_WP.pdf [April 2013].

Hansen, M., Hurwitz, W., and Madow, W. (1953). *Sample Survey Methods and Theory*. New York: John Wiley & Sons.

Hartley, H.O. (1962). Multiple frame surveys. Proceedings of the Social Statistics Section, American Statistical Association, 203-206.

Heaton, P. (2010). *Hidden in Plain Sight: What Cost-of-Crime Research Can Tell Us About Investing in Police*. Available: http://www.rand.org/content/dam/rand/pubs/occasional_papers/2010/RAND_OP279.pdf [March 2013].

Hodge, V., and Austin, J. (2004). A survey of outlier detection methodologies. *Artificial Intelligence Review, 22*(2), 85-126.

Holyk, G. (2008). Context effect. In P. Lavraka (Ed.), *Encyclopedia of Survey Research Methods* (p. 142). Thousand Oaks, CA: Sage.

Jabine, T., King, K., and Petroni, R. (1990). *Quality Profile for the Survey of Income and Program Participation (SIPP)*. Washington, DC: U.S. Census Bureau.

Jaquier, V., Johnson, H., and Fisher, B.S. (2011). Research methods, measures, and ethics. In C. Renzetti, J. Edleson, and R. Bergen (Eds.), *Source Book on Violence Against Women* (second ed., Ch. 2, pp. 23-45). Thousand Oaks, CA: Sage.

Kalsbeek, W., Spencer, B., and House, C. (2013). *The Relative Statistical and Operational Plausibility of Multiple-Frame Sampling for Rare Population Subgroups.* Paper presented at the Joint Statistical Meetings, Montreal, Quebec, Canada, August 6.

Kalton, G., Winglee, M., Krawchuk, S., and Levine, D. (2000). *Quality Profile for SASS: Rounds 1-3: 1987-1995.* (NCES #2000-308.) Washington, DC: National Center for Education Statistics, U.S. Department of Education.

Karjane, H., Fisher, B.S., and Cullen, F. (2005). *Sexual Assault on Campus: What Colleges and Universities Are Doing About It.* Available: https://www.ncjrs.gov/pdffiles1/nij/205521.pdf [March 2013].

Kasprzyk, D., and Kalton, G. (2001, May). *Quality Profiles in U.S. Statistical Agencies.* Paper presented at the International Conference on Quality in Official Statistics, Stockholm, Sweden.

Kilpatrick, D.G., Edmunds, C.N., and Seymour, A. (1992). *Rape in America: A Report to the Nation.* Charleston: Medical School University of South Carolina, National Victim Center and Crime Victims Research and Treatment Center.

Kilpatrick, D.G., Resnick, H.S., Ruggiero, K.J., Conoscenti, L.M., and McCauley, J. (2007). *Drug-Facilitated, Incapacitated, and Forcible Rape: A National Study.* Washington, DC: U.S. Department of Justice, National Institute of Justice. Available: http://www.ncbi.nlm.nih.gov/pmc/articles/PMC2967593 [October 2013].

Koss, M.P., Gidycz, C.A., and Wisniewski, N. (1987). The scope of rape: Incidence and prevalence of sexual aggression and victimization in a national sample of higher education students. *Journal of Counseling and Clinical Psychology, 55*(2), 162-170.

Koss, M.P., Abbey, A., Campbell, R., Cook, S., Norris, J., Testa, M., and White, J. (2007). Revising the SES: A collaborative process to improve assessment of sexual aggression and victimization. *Psychology of Women Quarterly, 31*(4), 357-370.

Krebs, C.P., Lindquist, C.H., Warner, T.D., Fisher, B.S., and Martin, S.L. (2007). *The Campus Sexual Assault (CSA) Study.* Washington, DC: U.S. Department of Justice, National Institute of Justice.

Lauritsen, J.L. (2012). *Subpopulations at High Risk for Rape and Sexual Assault: What Does the NCVS Tell Us?* Paper commissioned by the National Research Council Panel on Measuring Rape and Sexual Assault in the Bureau of Justice Statistics Household Surveys, Washington, DC. Available: http://sites.nationalacademies.org/DBASSE/CNSTAT/CurrentProjects/DBASSE_073316 [March 2013].

Lauritsen, J.L., and Heimer, K. (2008). The gender gap in violent crime victimization, 1973-2004. *Journal of Quantitative Criminology, 24*(2), 125-147.

Lauritsen, J.L., Owens, J., Planty, M., Rand, M., and Truman, J. (2012). *Methods for Counting High-Frequency Repeat Victimizations in the National Crime Victimization Survey.* Available: http: //bjs.gov/content/pub/pdf/mchfrv.pdf [April 2013].

Lizotte, A. (1985). Uniqueness of rape: Reporting assaultive violence to the police. *Crime and Delinquency, 31*(2), 169-190.

Lizotte, A., and Wolfson, F. (1981). *The Social Location of Rape and Reporting: A Victimization Study.* Paper presented at the annual meeting of the American Sociological Association, Montreal, Quebec, Canada.

Lohr, S. (2011). Alternative survey sample designs: Sampling with multiple overlapping frames. *Survey Methodology, 37*(2), 197-213.

Lynch, J., and Addington, L. (2007). *Understanding Crime Statistics: Revisiting the Divergence of the NCVS and UCR.* New York: Cambridge University Press.

Lynch, J., Berbaum, M., and Planty, M. (2002). *Investigating Repeated Victimization with the NCVS, Final Report.* Washington, DC: U.S. Department of Justice, National Institute of Justice. (Grant 97-IJ-CX-0027.) Available: https://www.ncjrs.gov/pdffiles1/nij/grants/193415.pdf [October 2013].

Maggi, L. (2009, July 11). NOPD downgrading of rape reports raises questions. *The Times-Picayune*. Available: http://s.nola.com/KJqWErP [March 2013].

Matza, M., McCoy, C., and Fazlollah, M. (1998, November 15). Pressure builds on city police for accuracy. *Philadelphia Inquirer*. Available: http://inquirer.philly.com/packages/crime/html/111598.asp [March 2013].

Meloy, M.L., and Miller, S.L. (2011). *The Victimization of Women: Law, Policies, and Politics*. New York: Oxford University Press.

National Research Council. (1976). *Surveying Crime*. Panel for the Evaluation of Crime Surveys. Committee on National Statistics. B. Penick and M. Owens, III (Eds.). Academy of Mathematical and Physical Sciences. Washington, DC: National Academy Press.

National Research Council. (2006). *Once and Only Once, and in the Right Place: Residence Rules in the Decennial Census*. Panel on Residence Rules in the Decennial Census. D. Cork and P. Voss (Eds.). Committee on National Statistics, Division of Behavioral and Social Sciences and Education. Washington, DC: The National Academies Press.

National Research Council. (2008). *Surveying Victims: Options for Conducting the National Crime Victimization Survey.* Panel to Review the Programs of the Bureau of Justice Statistics. R. Groves and D. Cork (Eds.). Committee on National Statistics and Committee on Law and Justice, Division of Behavioral and Social Sciences and Education. Washington, DC: The National Academies Press.

National Research Council. (2009). *Ensuring the Quality, Credibility, and Relevance of U.S. Justice Statistics*. Panel to Review the Programs of the Bureau of Justice Statistics. R. Groves and D. Cork (Eds.). Committee on National Statistics and Committee on Law and Justice, Division of Behavioral and Social Sciences and Education. Washington, DC: The National Academies Press.

National Research Council. (2011). *Measuring Rape and Sexual Assault in Bureau of Justice Statistics Household Surveys*. Available: http://www8.nationalacademies.org/cp/projectview.aspx?key=49418 [August 2013].

National Research Council. (2012). *Small Populations, Large Effects: Improving the Measurement of the Group Quarters Population in the American Community Survey*. Panel on Statistical Methods for Measuring the Group Quarters Population in the American Community Survey. P. Voss and K. Marton (Eds.). Committee on National Statistics, Division of Behavioral and Social Sciences and Education. Washington, DC: The National Academies Press.

National Research Council. (2013a). *Measuring What We Spend: Toward a New Consumer Expenditure Survey.* Panel on Redesigning the BLS Consumer Expenditure Surveys, D.A. Dillman and C.C. House (Eds). Committee on National Statistics, Division of Behavioral and Social Sciences and Education. Washington, DC: The National Academies Press.

National Research Council. (2013b). *Principles and Practices for a Federal Statistical Agency* (fifth edition). Committee on National Statistics. C. Citro and M. Straf (Eds.). Division of Behavioral and Social Sciences and Education. Washington, DC: The National Academies Press.

Neter, J., and Waksberg, J. (1964). A study of response errors in expenditures data from household interviews. *Journal of the American Statistical Association, 59*, 18-55.

NORC at the University of Chicago. (2009). *NORC Methodological Research to Support the Redesign of the National Crime Victimization Survey: Analysis of Possible Nonresponse Bias*. (Competitive ID #2008-BJS-1834D.) Chicago, IL: NORC at the University of Chicago.

Office for Victims of Crime. (n.d.). *Domestic and Family Violence*. Available: http://ovc.ncjrs.gov/topic.aspx?topicid=27 [March 2013].

Office of Applied Studies. (2003). NSDUH changes and their impact on trend measurement. In *Results from the 2002 National Survey on Drug Use and Health: National Findings* (Appendix C, pp. 107-137). Washington, DC: U.S. Department of Health and Human Services.

Pedlow, S., Porras, J., O'Muircheartaigh, C., and Shin, H. (2010). *Outlier Weight Adjustment in REACH 2010*. Proceedings of the Survey Research Methods Section, American Statistical Association. Available: http://www.amstat.org/sections/srms/Proceedings/y2003/Files/JSM2003-000939.pdf [October 2013].

Planty, M., and Strom, K. (2007). Understanding the role of repeat victims in the production of annual U.S. victimization rates. *Journal of Quantitative Criminology,* 23(3), 179-200.

Planty, M., Langton, L., Krebs, C., Berzofsky, M., and Smiley-McDonald, H. (2013). *Female Victims of Sexual Violence, 1994-2010*. (NCJ #240655.) Available: http://bjs.gov/content/pub/pdf/fvsv9410.pdf [April 2013].

President's Commission on Law Enforcement and the Administration of Justice. (1967). *The Challenge of Crime in a Free Society.* N. Katzenback, Chairman. (NCJ #000042.) Washington, DC: U.S. Government Printing Office.

Ramsey, C. (2010, September 14). *Rape in the United States: The Chronic Failure to Report and Investigate Rape Cases*. Hearing before the Senate Committee on the Judiciary, Subcommittee on Crime and Drugs. Available: http://www.judiciary.senate.gov/pdf/09-14-10%20Ramsey%20Testimony.pdf [April 2013].

Rand, M. (2009). *Redesigning the National Crime Victimization Survey*. Presentation at the Federal Committee on Statistical Methodology Research Conference, Washington, DC. Available: www.fcsm.gov/09papers/Rand_X-B.doc [September 2013].

Rand, M., and Rennison, C. (2005). Bigger is not necessarily better: An analysis of violence against women estimates from the National Crime Victimization Survey and the National Violence Against Women Survey. *Journal of Quantitative Criminology, 21*(3), 267-291.

Rape, Abuse and Incest National Network. (n.d.). *Preserving and Collecting Forensic Evidence*. Available: http://www.rainn.org/get-information/aftermath-of-sexual-assault/preserving-and-collecting-forensic-evidence [October 2013].

Rasinski, K. (2012). *Design and Implementation Strategies Used to Collect Information on Rape and Sexual Assault: A Review of the Literature with Recommendations for the NCVS*. Paper commissioned by the National Research Council Panel on Measuring Rape and Sexual Assault in the Bureau of Justice Statistics Household Surveys, Washington, DC. Available: http://sites.nationalacademies.org/DBASSE/CNSTAT/CurrentProjects/DBASSE_073316#.UVSWS1tmtJE [March 2013].

Rennison, C.M., and Rand, M. (2007). Introduction to the National Crime Victimization Survey. In J. Lynch and L. Addington (Eds.), *Understanding Crime Statistics: Revisiting the Divergence of the NCVS and UCR* (pp. 17-54). New York: Cambridge University Press.

Resnick, H., Kilpatrick, D., Dansky, B., Saunders, B., and Best, C. (1993). Prevalence of civilian trauma and PTSD in a representative national sample of women. *Journal of Consulting and Clinical Psychology, 61*(6), 984-991.

Schaeffer, N.C. (2000). Asking questions about threatening topics: A selective overview. In A. Stone, J. Turkkan, C. Bachrach, J. Jobe, H. Kurtzman, and V. Cain (Eds.), *The Science of Self-Report: Implications for Research and Practice* (pp. 105-122). Mahwah, NJ: Lawrence Erlbaum Associates.

Schreck, C., Stewart, E., and Fisher, B. (2006). Self-control, victimization, and their influence on risky lifestyles: A longitudinal analysis using panel data. *Journal of Quantitative Criminology,* 22(4), 319-340.

Schwarz, N. (1996). *Cognition and Communication: Judgmental Biases, Research Methods, and the Logic of Conversation.* Hillsdale, NJ: Lawrence Erlbaum Associates.

Strack, F., and Martin, L. (1987). Thinking, judging, and communicating: A process account of context effects in attitude surveys. In H.J. Hippler, N. Scharz, and S. Sudman (Eds.), *Social Information Processing and Survey Methodology* (pp. 123-148). New York: Springer-Verlag.

Sudman, S., and Bradburn, N. (1982). *Asking Questions.* San Francisco, CA: Jossey-Bass.

Thissen, R., Fisher, C., Barber, L., and Sattaluri, S. (2009). *Computer Audio-Recorded Interviewing (CARI): A Tool for Monitoring Field Interviewers and Improving Field Data Collection.* Available: http://www.statcan.gc.ca/pub/11-522-x/2008000/article/10955-eng.pdf [October 2013].

Tjaden, P., and Thoennes, N. (2000). *Full Report of the Prevalence, Incidence, and Consequences of Violence Against Women: Findings from the National Violence Against Women Survey.* (NCJ #183781.) Washington, DC: U.S. Department of Justice, National Institute of Justice.

Tjaden, P., and Thoennes, N. (2006). *Extent, Nature, and Consequences of Rape Victimization: Findings from the National Violence Against Women Survey.* (NCJ #210346.) Washington DC: U.S. Department of Justice, National Institute of Justice.

Tourangeau, R. (1984). Cognitive science and survey methods: A cognitive perspective. In T. Jabine, M. Straf, J. Tanur, and R. Tourangeau (Eds.), *Cognitive Aspects of Survey Methodology: Building a Bridge Between Disciplines* (pp. 73-100). Washington, DC: National Academy Press.

Tourangeau, R., and Smith, T. (1996). Asking sensitive questions: The impact of data collection mode, question format, and question context. *Public Opinion Quarterly, 60*(2), 275-304.

Tourangeau, R., and Yan, T. (2007). Sensitive questions in surveys. *Psychological Bulletin, 133*(5), 859-883.

Tourangeau, R., Rips, L., and Rasinski, K. (2000). *The Psychology of Survey Response.* Cambridge, UK: Cambridge University Press.

Tracy, C., Fromson, T., Long, J., and Whitman, C. (2012). *Rape and Sexual Assault in the Legal System.* Paper commissioned by the National Research Council Panel on Measuring Rape and Sexual Assault in the Bureau of Justice Statistics Household Surveys. Available: http://sites.nationalacademies.org/DBASSE/CNSTAT/CurrentProjects/DBASSE_073316 [March 2013].

U.S. Census Bureau. (2000). *Historical Census of Housing Tables—Telephones.* Available: http://www.census.gov/hhes/www/housing/census/historic/phone.html [April 2013].

U.S. Census Bureau. (2006). *Design and Methodology: Current Population Survey.* (Technical Paper #66.) Available: http://www.census.gov/prod/2006pubs/tp-66.pdf [April 2013].

U.S. Census Bureau. (2010). *American Community Survey/Puerto Rico Community Survey Group Quarters Definitions.* Available: http://www.census.gov/acs/www/Downloads/data_documentation/GroupDefinitions/2010GQ_Definitions.pdf [March 2013].

U.S. Census Bureau and Bureau of Justice Statistics. (2008). *NCVS CAPI Interviewing Manual for Field Representatives.* Available: http://www.reginfo.gov/public/do/PRAViewIC?ref_nbr=201206-1121-001&icID=12401 [May 2013].

U.S. Census Bureau and Bureau of Justice Statistics. (2010). *National Crime Victimization Survey (NCVS) CAPI Reinterviewer's Manual.* Available: http://www.reginfo.gov/public/do/PRAViewIC?ref_nbr=201206-1121-001&icID=12401 [October 2013].

U.S. Census Bureau and Bureau of Justice Statistics. (2011a). *NCVS CAPI Refresher Training Guide: A Renewed Focus on Data Quality.* Available: http://www.reginfo.gov/public/do/PRAViewIC?ref_nbr=201206-1121-001&icID=12401 [May 2013].

U.S. Census Bureau and Bureau of Justice Statistics. (2011b). *NCVS CAPI Refresher Training Workbook: A Renewed Focus on Data Quality.* Available: http://www.reginfo.gov/public/do/PRAViewIC?ref_nbr=201206-1121-001&icID=12401 [May 2013].

U.S. Energy Information Administration. (1996). *Residential Energy Consumption Survey Quality Profile*. Washington, DC: U.S. Department of Energy.

Weiss, K.G. (2010). Too ashamed to report: Deconstructing the shame of sexual victimization. *Feminist Criminology, 5*(3), 286-310.

Williams, C. (2012, March 18). Buried amid rape kit backlog: Justice for victims. *Detroit Free Press*. Available: http://www.freep.com/apps/pbcs.dll/article?AID=/201203182355/NEWS01/120318040 [March 2013].

Wolfgang, M.E. (1963). Uniform Crime Reports: A critical appraisal. *University of Pennsylvania Law Review, 111*(6), 708-738.

Wolitzky-Taylor, K., Resnick, H., McCauley, J., Amstadter, A., Kilpatrick, D., and Ruggiero, K. (2011). Is reporting of rape on the rise? A comparison of women with reported versus unreported rape experiences in the National Women's Study-Replication. *Journal of Interpersonal Violence, 26*(4), 807-832.

Yu, Q., Stasny, E., and Lin, B. (2008). Bayesian models to adjust for response bias in survey data for estimating rape and domestic violence rates from the NCVS. *The Annals of Applied Statistics, 2*(2), 665-686.

Appendix A

Need for the Study[1]

Self-report surveys of criminal victimization were a breakthrough in crime statistics and are acknowledged as an important part of any national statistical system on crime and criminal justice. It is essential that the police and the criminal justice system not be the only source of data on crime and responses to crime. Surveys give citizens a direct voice in the definition of the crime problem. This is particularly important in the case of measuring rape and sexual assault, because there is good evidence that the majority of these offenses are not reported to the police. These offenses remain the darkest of the "dark figure" of crime.

The greater acceptance of the self-report method has resulted in the fielding of a variety of surveys employing a wide range of methodologies. This blossoming of surveys has led to the recognition that the methods employed in asking about victimization can have a substantial impact on the volume and nature of that behavior reported in the survey. While having a variety of methods provides important information on crime, varying results have raised questions about the suitability of specific surveys and about the self-report method more generally.

In the case of estimates of rape and sexual assault from self-report surveys, two schools of thought have emerged with somewhat different goals and very different methodologies. One group emphasizes the criminal justice perspective, and the other takes a public health approach. The crimi-

[1]James Lynch, director of the Bureau of Justice Statistics (the study sponsor), presented comments to the panel at its first meeting on December 8, 2011. Lynch provided the written version in this appendix to the panel the following week.

nal justice school emphasizes crime as a point-in-time event and employs legal definitions (but plain language descriptions) of the target behavior. As a result, the survey methods used emphasize placing an event in time, collecting an extensive amount of information about that event, and using that information to determine whether the reported event satisfies the legal definitions of victimization. The National Crime Victimization Survey (NCVS) is the icon of this approach and has introduced specific procedures that have become identified with the criminal justice school.

The public health approach emphasizes victimization as a condition that endures over time and requires treatment to restore the victim. Consequently, there is less concern with identifying point-in-time events that may comprise the condition, and legal definitions are of less concern than commonly understood definitions of the behavior. Issues of coercion, consent, and complicity that are so central to the definition of a criminal act are not asked about in the public health tradition. The survey methods employed reflect this orientation. Explicit and extensive cues are used to prompt mention of the conditions of interest. Very little attention is paid to situating events in time or collecting extensive information on the event to determine if it satisfies the condition for inclusion. More attention is given to the consequences of the victimization, its duration, and its social context. There are a number of surveys that have taken this approach in varying degrees including the National Woman's Study (NWS), National Violence Against Women Study (NVAWS), and, more recently, the National Intimate Partner and Sexual Violence Surveillance System (NISVSS).

Although these two schools of thought are different in goals and in method, there is considerable overlap and potential complementarity between them. The public health tradition, for example, has led the way in strategies for stimulating the recall and reporting of rape and sexual assault. The criminal justice tradition has pioneered methods for situating events in time and filtering out the ineligible. The discourse between the two groups, however, has been largely defensive with the result that very little light has been shed on the problem of measuring rape and sexual assault. Our hope in sponsoring this panel is that a group of substantive and methodological experts can take a fresh look at the problem, drawing from what the criminal justice and public health schools have done but not being held captive by these traditions. The principal goal of the panel is to consider a wide range of alternative self-report survey designs to measure the incidence and prevalence of the crimes of rape and sexual assault and to recommend an optimum design.

A second charge to the panel is to recommend whether this optimum design can be incorporated into the ongoing NCVS program and, if so, how. The optimum design may only be able to be implemented as a free-standing survey that would be administered at fixed intervals and used to adjust annual estimates from the core NCVS. Alternatively the design may be able to be fielded as a supplement to the core NCVS or even as part of the core survey.

The evaluation of different designs should take account of the mission of BJS. The panel should be mindful that BJS is responsible for providing estimates of the incidence and prevalence of crimes, and any design recommended must be optimum relative to measuring behavior defined by the law as criminal. In screening for the target behavior, however, broader definitions of the target events may be used in the screening process, but ultimately criminal behavior must be identifiable. The principal population of interest is the noninstitutionalized residential population of the United States. The panel may consider age limits on the target population as survey procedures dictate. Other populations can be accommodated in the optimum design as long as their inclusion does not adversely affect estimates for this principal population or have a large impact on cost. The most important estimates to be obtained from the survey are national-level and change estimates for a specified unit of time. These estimates are designed to be interpreted as risk rates. Annual estimates are typical, but other reference and reporting periods can be considered if appropriate. Change estimates need not be based on consecutive years. The survey should also provide detailed information on the victimization incident, the sequelae of victimization, and the criminal justice and treatment responses.

Finally, the panel is asked to work closely with Westat in field testing the recommended design. Ideally, the panel's deliberations would be both complete and vetted before a field test would be undertaken, but because of uncertainty regarding funding, the panel's work and the field test must proceed almost simultaneously. We ask the panel to share its recommendations with Westat and BJS as soon as prudence and the requirements of the deliberation process allow. Westat will proceed with work on the companion design as the panel deliberates. BJS and Westat will incorporate the guidance of the panel into the implementation of the optimum design as the recommendations emerge.[2]

[2]As requested by Director Lynch, the panel worked publicly with investigators at Westat, which had been contracted by BJS to develop a pilot project to test two alternative survey designs to measure rape and sexual assault. Westat staff presented the status of their work at each of the panel's open meetings and participated in open discussion at those meetings with panel members and other participants. Following the June 5-6, 2012, public workshop, several panel members provided individual informal written feedback to Westat on the draft plans that Westat presented at that public workshop. These were not consensus conclusions of the panel. They are provided for the purpose of full disclosure in the Public Assess File. The Westat team and the panel kept each other advised of their project timelines for various activities throughout the process. Following National Research Council policy, there was no sharing of the panel's deliberations, conclusions, or recommendations with Westat or BJS during the study.

Appendix B

Workshop and Public Meetings: Agendas and Participants

First Meeting of the Panel on Measuring Rape and Sexual Assault in Bureau of Justice Statistics Household Surveys

Thursday, December 8, 2011
Keck Center
500 Fifth Street NW, Room 110
Washington, DC 20001

AGENDA

OPEN SESSION (open to public)
8:00 a.m.-5:15 p.m.

	Session Chair AM: William Kalsbeek *Session Chair PM:* Candace Kruttschnitt
8:00-8:30 a.m.	**Networking with Continental Breakfast**
8:30-9:00 a.m.	**Welcome and Introductions** William Kalsbeek, *Cochair* Candace Kruttschnitt, *Cochair* Connie Citro, *Committee on National Statistics* Robert Hauser, *Division of Behavioral and Social Sciences and Education*

9:00-9:30 a.m.	**Need for the Study** James Lynch *Bureau of Justice Statistics*
9:30-9:45 a.m.	**Break**
9:45-10:15 a.m.	**Police Reports on Rape and Sexual Assault** Lynn Addington, *Department of Justice, Law and Society, American University*
10:15 a.m.-12:15 p.m.	**Self-Report Methodology: A Criminal Justice Approach** National Crime Victimization Survey: Its History and General Design Michael Rand, *Bureau of Justice Statistics* Overview of NCVS Redesign Efforts Allen Beck, *Bureau of Justice Statistics*
12:15-1:15 p.m.	**Working Lunch** (third floor atrium—lunch tickets provided)
1:15-2:30 p.m.	**Self-Report Methodology: Alternative Approaches** National Violence Against Women Study Patricia Tjaden, *Tjaden Research Corporation* National Intimate Partner and Sexual Violence Survey (NISVS) E. Lynn Jenkins, *Division of Violence Prevention, National Center for Injury Prevention and Control*
2:30-2:45 p.m.	**Break** (refreshments available in room)

2:45-3:45 p.m.	**View from Stakeholders**
	Laura Dugan, *Department of Criminology and Criminal Justice, University of Maryland, College Park*
	Scott Berkowitz, *RAINN (Rape, Abuse and Incest National Network)*
	Carol Tracy, *Women's Law Project*
3:45-4:15 p.m.	**Self-Report Methodology: Experiences in Implementing the Prison Rape Elimination Act** Allen Beck, *Bureau of Justice Statistics*
4:15-4:45 p.m.	**BJS-Sponsored Collaboration with Westat to Develop Alternative Approaches** David Cantor, *Westat*
4:45- 5:15 p.m.	**Public Input and Open Discussion on the Focus of the Committee** Candace Kruttschnitt, *Facilitator*
5:15 p.m.	**Adjournment of Public Session**

Workshop on Measuring
Rape and Sexual Assault in Bureau of Justice Stastics Surveys

June 5-6, 2012
National Academy of Sciences
2101 Constitution Avenue, Room 120
Washington, DC 20418

AGENDA

Day 1: Tuesday, June 5, 2012, Main Building, Conference Room 120
OPEN MEETING (open to public)
8:30 a.m.-5:15 p.m.

8:30-9:00 a.m. **Networking and Continental Breakfast** (inside room 120)

9:00- 9:30 a.m. **Welcome, Introductions, Work of Panel and Purpose of Workshop**
William Kalsbeek and Candace Kruttschnitt, *Panel Cochairs*

Session 1A: The Target—What Are We Trying to Measure?
Facilitator: William Kalsbeek

9:30-10:00 a.m. **BJS Program Goals and Expectations**
James Lynch, *Bureau of Justice Statistics*
Allen Beck, *Bureau of Justice Statistics*

10:00-10:30 a.m. **Discussion**
Discussion Leader: John Boyle, *Abt SRBI*

10:30-11:00 a.m. **Break** (refreshments available inside room 120)

11:00-11:30 a.m. **Defining Rape and Sexual Assault in the Legal System**
Carol Tracy, *Women's Law Center*
Jennifer Long, *AEquitas*

11:30 a.m.-12:00 p.m. **Discussion**
Discussion Leader: Ruth Peterson, *Ohio State University*

12:00-1:00 p.m. **Working Lunch to Discuss Afternoon Sessions** (main dining room; lunch tickets will be provided)

Session 2: The Landscape—Where Are We Now?
Facilitator: Candace Kruttschnitt

1:00-1:30 p.m. **An Examination of Ignorable Nonresponse in the National Crime Victimization Survey**
Fritz Scheuren, *NORC at the University of Chicago*

1:30-2:00 p.m. **Discussion**
Discussion Leader: Nora Cate Schaeffer, *University of Wisconsin–Madison*

2:00-2:30 p.m. **Subpopulations at High Risk for Rape and Sexual Assault: What Does the NCVS Tell Us?**
Janet Lauritsen, *University of Missouri, St. Louis*

2:30-3:00 p.m. **Discussion**
Discussion Leader: Colin Loftin, *University at Albany*

3:00-3:30 p.m. **Break** (refreshments available in room 120)

Session 3A: The Tools—What New Measurement Tools Will Be Helpful?
Facilitator: William Kalsbeek

3:30-4:00 p.m. **Alternative Survey Designs and Implementation Strategies Used to Collect Sensitive Data**
Ken Rasinski, *University of Chicago*

4:00-4:30 p.m. **Discussion**
Discussion Leader: Carol House, *Project Study Director*

4:30-5:00 p.m. **Wrap-Up**
William Kalsbeek and Candace Kruttschnitt, *Panel Cochairs*

5:00 p.m. **Adjournment**

Day 2: Wednesday, June 6, 2012, NAS Building, Conference Room 120
OPEN MEETING (open to public)
8:30 a.m.-4:00 p.m.

8:30-9:00 a.m.	**Networking and Continental Breakfast** (inside room 120)

Session 1B: The Target—What Are We Trying to Measure? Redux
Facilitator: William Kalsbeek

9:00-9:30 a.m.	**Defining Rape and Sexual Assault in Surveys and Other Sources** Ronet Bachman, *University of Delaware*
9:30-10:00 a.m.	**Discussion** *Discussion Leader:* Karen Heimer, *University of Iowa*
10:00-10:30 a.m.	**Break** (refreshments available in room 120)

Session 3B: The Tools—What New Measurement Tools Will Be Helpful? (Continued)
Facilitator: William Kalsbeek

10:30-11:00 a.m.	**Statistical Methods for Assessing the Bias in Estimates of Rates of Sensitive, Rare Events** Marcus Berzofsky, *RTI International*
11:00-11:30 a.m.	**Discussion** *Discussion Leader:* Paul Biemer, *RTI International*
11:30 a.m.-12:00 p.m.	**Multiple Frame Surveys** David Wilson and Jim Chromy, *RTI International*
12:00-12:30 p.m.	**Discussion** *Discussion Leader:* Tom Smith, *NORC at the University of Chicago*
12:30-12:45 p.m.	**Wrap-Up** William Kalsbeek and Candace Kruttschnitt, *Panel Cochairs*

12:45 p.m.	**Adjournment for Workshop**
12:45-2:00 p.m.	**Working Lunch to Discuss Afternoon Sessions** (main dining room; lunch tickets will be provided)
2:00-4:00 p.m.	**Roundtable Collaboration: Discussion of Workshop, Westat's Progress, and Panel's Work (informal discussion open to the public)**
2:00-4:00 p.m.	**Informal Discussion** *Facilitator:* Candace Kruttschnitt

Topics:

- Continuing discussion of workshop topics
- What "constitutes" a crime on NCVS? *Discussion Leader:* Allen Beck and others from BJS
- Panel's ideas with feedback from others *Discussion Leader:* William Kalsbeek
- Westat's progress, insights, and timeline *Discussion Leaders:* David Cantor (Westat) and Allen Beck
- Other topics

4:00 p.m.	**Adjournment**

Panel on Measuring Rape and Sexual Assault on BJS Household Surveys
Closed Meeting–Fourth Meeting of Panel

December 10-11, 2012
Keck Center
500 F Street, NW, Room 205
Washington, DC 20001

December 10, 2012—CLOSED MEETING

December 11, 2012—OPEN MEETING, PUBLIC INVITED

10:15 a.m.-12:15 p.m.	OPEN SESSION Discussion with Westat and BJS About Parallel Project

Appendix C

Links to Questionnaires of the National Crime Victimization Survey

Three survey instruments support the National Crime Victimization Survey: the Control Card, the Basic Screen Questionnaire, and the Incident Report. These survey instruments can be found on the website of the Bureau of Justice Statistics.

Control Card:
http://www.bjs.gov/content/pub/pdf/ncvs500_2011.pdf [September 2013]

Basic Screen Questionnaire:
http://www.bjs.gov/content/pub/pdf/ncvs109.pdf [September 2013]

Incident Report:
http://www.bjs.gov/content/pub/pdf/ncvs209.pdf [September 2013]

Appendix D

Selected Surveys Measuring Rape: An Overview

This appendix presents details, compiled by the panel, for eight surveys that have measured rape. Five of the surveys are discussed in Chapters 5 and 6:

- National Crime Victimization Survey
- National Women's Study
- National Violence Against Women Study
- National College Women Sexual Victimization Study
- National Intimate Partner and Sexual Violence Survey

The other three surveys are not discussed in the report:

- National Survey of Inter-Gender Relationships
- National Women's Study: Replicate
- Campus Sexual Assault Study

The material for each begins with a table that details the design and estimate(s). For most of the surveys, a second table shows the concepts and descriptions used in the survey. These one or two tables are followed by selected other information, such as the questions used.

TABLE D-1 National Crime Victimization Survey: Design and Estimates

Design		Data Collection	
Research Design	Population/ Sampling Design, N (age range)	Mode of Administration	Framing of Survey Context
Panel: Every 6 months for up to seven interviews over a 3-year period	Housing units in the United States/ stratified, multistage cluster design $n \cong 80{,}000$ households for each year $n \cong 145{,}000$ persons for each year* (All household members ages 12 and older) Proxy interviews included	Wave 1 interview is face to face; remaining ones are through decentralized telephone interview whenever feasible First interview used for bounding purposes until 2006; since that time, the Census Bureau has included the first interview in the estimates with a special adjustment for potential telescoping	Criminal victimization

*Number of households and persons interviewed changes slightly from year to year.
SOURCE: Data from Bureau of Justice Statistics (2008b).

NATIONAL CRIME VICTIMIZATION SURVEY

The National Crime Victimization Survey is an ongoing survey of the Bureau of Justice Statistics (see Chapter 4). Table D-1 presents details about its design and estimates and Table D-2 shows the survey's concept and its description. Following these two basic tables are excerpts from the victimization and incident screen questions.

Forms of Sexual Victimization	Operationalization			
Type(s) Definitions	Measurement Process (reference frame)	Screen Question(s)	Incident Question(s)	Estimate
Rape: completed, attempted, verbal threat Sexual assault: including verbal threats	Two stages: (1) victimization screening questions and (2) incident report (In past 6 months) Hierarchical scoring procedure used in incident report to classify type of victimization, if any, that occurred	Multiple items with short cues used	Responses from multiple items used	Annual rates per 1,000 persons ages 12 and older

TABLE D-2 National Crime Victimization Survey: Concept and Description

Concept	Description
Rape and attempted rape	**Rape** includes psychological coercion as well as physical force. Forced sexual intercourse means vaginal, anal, or oral penetration by the offender(s). It also includes incidents where the penetration is from a foreign object. It includes attempted rapes, male as well as female victims, and both heterosexual and homosexual rape. **Attempted rape** includes verbal threats of rape.
Sexual assault	**Sexual assault** included in this category includes a wide range of victimizations, separate from rape or attempted rape. These crimes include attacks or attempted attacks generally involving unwanted sexual contact between victim and offender. Sexual assaults may or may not involve force and include such things as grabbing or fondling. Sexual assault also includes verbal threats.

SOURCE: Data from Bureau of Justice Statistics (n.d.-b).

Example of Victimization Screen Question and Incident Screen Questions

Incidents involving forced or unwanted sexual acts are often difficult to talk about. (Other than any incidents already mentioned,) have you been forced or coerced to engage in unwanted sexual activity by - **(a) Someone you didn't know before -** **(b) A casual acquaintance -** **OR** **(c) Someone you know well?** ASK only if necessary **Did any incidents of this type happen to you?**	545 1 ☐ Yes - ASK 43b 2 ☐ No - SKIP to 44a
43b. **How many times?**	546 ________ Number of times (43b)
43c. **What happened?**	Briefly describe incident(s) ________ ________ ________

SOURCE: Data from Bureau of Justice Statistics (n.d.-d).

28e. **You mentioned some type of unwanted sexual contact with force. Do you mean forced or coerced sexual intercourse including attempts?**	1 ☐ Yes - ASK 29a 2 ☐ No - SKIP to 39b
29a. **How were you attacked?** Probe: **Any other way?** Enter all that apply.	546 * 1 ☐ Raped .... - ASK 29c 2 ☐ Tried to rape .... - ASK 29d 3 ☐ Sexual assault other than rape or attempted rape 4 ☐ Shot 5 ☐ Shot at (but missed) 6 ☐ Hit with gun held in hand 7 ☐ Stabbed/cut with knife/sharp weapon 547 * 8 ☐ Attempted attack with knife/sharp weapon 9 ☐ Hit by object (other than gun) held in hand 10 ☐ Hit by thrown object 548 * 11 ☐ Attempted attack with weapon other than gun/knife/sharp weapon 12 ☐ Hit, slapped, knocked down 13 ☐ Grabbed, held, tripped, jumped, pushed, etc. (3–13: SKIP to 30a) 14 ☐ Other - Specify - ASK 29b
29b. Please specify how were you attacked.	Specify - SKIP to 30a ________
29c. **You mentioned rape. Do you mean forced or coerced sexual intercourse?** If "no", then ask: **What do you mean?**	☐ Yes - SKIP to 30a ☐ No - go back to 29a
29d. **You mentioned attempted rape. Do you mean attempted forced or coerced sexual intercourse?** If "no", then ask: **What do you mean?** Page 6	☐ Yes - SKIP to 30a ☐ No - go back to 29a

SOURCE: Data from Bureau of Justice Statistics (n.d.-d).

30a. **Did the offender THREATEN to hurt you before you were actually attacked?**	649 1 ☐ Yes } SKIP to 31a 2 ☐ No } 3 ☐ Other - Specify - ASK 30b
30b. Please specify	Specify ________
31a. **What were the injuries you suffered, if any?** Probe: **Anything else?** Enter all that apply.	655 * 1 ☐ None .... SKIP to 39b 2 ☐ Raped .... SKIP to 31c 3 ☐ Attempted rape .... SKIP to 31d 656 * 4 ☐ Sexual assault other than rape or attempted rape 5 ☐ Knife or stab wounds 6 ☐ Gun shot, bullet wounds 7 ☐ Broken bones or teeth knocked out 8 ☐ Internal injuries 9 ☐ Knocked unconscious 10 ☐ Bruises, black eye, cuts, scratches, swelling, chipped teeth (4–10: SKIP to 32) 11 ☐ Other - Specify - ASK 31b
31b. Please specify the injuries you suffered.	Specify ________ - SKIP to 32
31c. **You mentioned rape. Do you mean forced or coerced sexual intercourse?** If "no", then ask: **What do you mean?**	☐ Yes - SKIP to 32 ☐ No - go back to 31a

SOURCE: Data from Bureau of Justice Statistics (n.d.-d).

31d. **You mentioned attempted rape. Do you mean attempted forced or coerced sexual intercourse?** If "no", then ask: **What do you mean?**	☐ Yes - SKIP to 32 ☐ No - go back to 31a
32. Ask or verify- **Were any of the injuries caused by a weapon other than a gun or knife?**	657 1 ☐ Yes - ASK 33 2 ☐ No - SKIP to 34
33. **Which injuries were caused by a weapon OTHER than a gun or knife?** Enter all that apply.	658 * 2 ☐ Raped 3 ☐ Attempted rape 4 ☐ Sexual assault other than rape or attempted rape 7 ☐ Broken bones or teeth knocked out 8 ☐ Internal injuries 9 ☐ Knocked unconscious 10 ☐ Bruises, black eye, cuts, scratches, swelling, chipped teeth 11 ☐ Other - Specify
34. **Were you injured to the extent that you received any medical care, including self treatment?**	659 1 ☐ Yes - ASK 35a 2 ☐ No - SKIP to 39b
35a. **Where did you receive this care?** Probe: **Anywhere else?** Enter all that apply.	660 * 1 ☐ At the scene .... 2 ☐ At home/neighbor's/friend's .... 3 ☐ Health unit at work/school, first aid station at a stadium/park, etc. .... 4 ☐ Doctor's office/health clinic .... 5 ☐ Emergency room at hospital/emergency clinic .... 6 ☐ Hospital (other than emergency room) .... (1–6: SKIP to 35c) 7 ☐ Other - Specify - ASK 35b
35b. Please specify where you received this care.	Specify ________

SOURCE: Data from Bureau of Justice Statistics (n.d.-d).

TABLE D-3 National Women's Study: Design and Estimates

Design		Data Collection	
Research Design	Population/ Sampling Design, Sample Size, Age Range	Mode of Administration	Framing of Survey Context
3-year longitudinal	Probability sample Two stages: (1) geographic regions; (2) Random digit dialing to select households within each area *n* = 2,008 a cross-section of all adult women (ages 18 and older) *n* = 2,000 an over sample of younger women between the ages of 18 and 34	Telephone interview Wave 1: Initial interview Wave 2: 1-year follow-up Wave 3: 2-year follow-up	Preamble to questions: "Women do not always report such experiences to police or discuss them with family or friends."

SOURCE: Data from Kilpatrick, Edmunds, and Seymour (1992).

NATIONAL WOMEN'S STUDY

The National Women's Study, supported by the National Institute of Drug Abuse, was conducted in 1989-1990 (see Chapter 5). Table D-3 presents details about its design and estimates and Table D-4 shows the survey's concept and its description. They are followed by the questions used in the survey.

Forms of Sexual Victimization	Operationalization			
Type(s)	Measurement Process (reference frame)	Screen Question(s)	Incident Question(s)	Estimate
Forceful rape	Single stage Behaviorally specific (Lifetime prevalence of rape) (The past year prevalence of rape)	Four single items		Percentage Estimated counts

TABLE D-4 National Women's Study: Concept and Description

Concept	Description
Forcible rape	**Rape** was defined as "an event that occurred without the woman's consent which involved the use of force or threat of force, and involved sexual penetration of victim's vagina, mouth or rectum." **The critical elements of forcible rape:** 1. use of force or threat of force 2. lack of consent, and 3. sexual penetration.

SOURCE: Kilpatrick, Edmunds, and Seymour (1992).

Survey Questions

This study was designed to ask American women provocative, personal questions in order to leave no doubt or confusion as to the definition of forcible rape. The questions themselves were difficult to ask—and equally difficult for women to answer—but they provide clear answers for the first time to the critical elements of forcible rape:

- Use of force or threat of force;
- Lack of consent; and
- Sexual penetration.

Here are questions asked in *The National Women's Survey*:

"...***Women do not always report such experiences to police or discuss them with family or friends. The person making the advances isn't always a stranger, but can be a friend, boyfriend, or even a family member. Such experiences can occur anytime in a woman's life—even as a child. Regardless of how long ago it happened or who made the advances...***"

- Has a man or boy ever made you have sex by using force or threatening to harm you or someone close to you? Just so there is no mistake, by sex we mean putting a penis in your vagina.
- Has anyone ever made you have oral sex by force or threat of harm? Just so there is no mistake, by oral sex, we mean that a man or boy put his penis in your mouth or somebody penetrated your vagina or anus with his mouth or tongue.
- Has anyone ever made you have anal sex by force or threat of harm?
- Has anyone ever put fingers or objects in your vagina or anus against your will by using force or threat?"

SOURCE: Data from Kilpatrick, Edmunds, and Seymour (1992).

NATIONAL VIOLENCE AGAINST WOMEN STUDY

The National Violence Against Women Study was mandated by Congress in the 1994 Violence Against Women Act (see Chapter 5). It was conducted in 1995 under the joint sponsorship of the National Institute of Justice and the Centers for Disease Control and Prevention. Table D-5 presents details about its design and estimates and Table D-6 shows the survey's concept and its description. They are followed by the questions used in the survey.

TABLE D-5 National Violence Against Women Study: Design and Estimates

Design		Data Collection	
Research Design	Population/ Sampling Design, Sample Size, Age Range	Mode of Administration	Framing of Survey Context
Cross-section	Probability sample Two stages: (1) census regions; (2) random digit dialing to select households within each area *n* = 8,000 a cross-section of all adult women (ages 18 and older) *n* = 8,005 a cross-section of all adult men (ages 18 and older)	Telephone interview using computer-assisted telephone interviewing; used all female interviewers	Personal safety

SOURCE: Data from Tjaden and Thoennes (2000).

TABLE D-6 National Violence Against Women Study: Concept and Description

Concept	Description
Forcible rape	**Rape** was defined as "an event that occurred without the victim's consent, which involved the use of force or threat of force, and involved sexual penetration of victim's vagina, or anus by penis, tongue, fingers, or object, or the victim's mouth by penis. The definition included both attempted and completed rape." **The critical elements of forcible rape:** 1. use of force or threat of force, 2. lack of consent, and 3. sexual penetration.

SOURCE: Data from Tjaden and Thoennes (2000).

Form of Sexual Victimization	Operationalization			
Type	Measurement Process (reference frame)	Screen Question(s)	Incident Question(s)	Estimate
Completed and attempted forcible rape	Single stage Behaviorally specific (Lifetime prevalence of rape) (The past year prevalence of rape) (12-month incidence rates)	Five single items	Single-stage classification process, with no separate incident report	Percentage Estimated counts

SURVEY QUESTIONS:

- [Female respondents only] *Has a man or boy ever made you have sex by using force or threatening to harm you or someone close to you? Just so there is no mistake, by sex we mean putting a penis in your vagina.*
- *Has anyone, male or female, ever made you have oral sex by using force or threat of force? Just so there is no mistake, by oral sex we mean that a man or boy put his penis in your mouth or someone, male or female, penetrated your vagina or anus with their mouth.*
- *Has anyone ever made you have anal sex by using force or threat of harm? Just so there is no mistake, by anal sex we mean that a man or boy put his penis in your anus.*
- *Has anyone, male or female, ever put fingers or objects in your vagina or anus against your will or by using force or threats?*
- *Has anyone, male or female, ever* ***attempted*** *to make you have vaginal, oral, or anal sex against your will but intercourse or penetration did not occur?*

TABLE D-7 National College Women Sexual Victimization Study: Design and Estimates

Design		Data Collection	
Research Design	Population/ Sampling Design, Sample Size, Age Range	Mode of Administration	Framing of Survey Context
Cross-section	2-year and 4-year colleges and universities Two stages: (1) stratified institutions by total student enrollment and location of school; (2) randomly selected women enrolled in selected institutions Number of institutions, 233 n = 4,446	Computer-assisted telephone interviewing system	Unwanted sexual experiences that women may encounter during college

SOURCE: Data from Fisher et al. (2000).

Other questions were included regarding physical assault, stalking, victim-perpetrator relationship, and the characteristics and consequences of violence.

SOURCE: Data from Tjaden and Thoennes (2000).

NATIONAL COLLEGE WOMEN SEXUAL VICTIMIZATION STUDY

The National College Women Sexual Victimization Study was conducted in 1997, supported by the National Institute of Justice (see Chapter 5). Table D-7 presents details about its design and estimates and Table D-8 shows the survey's concept and its definitions. They are followed by the detailed questionnaire used in the survey.

Form of Sexual Victimization	Operationalization			
Type(s)	Measurement Process (reference frame)	Screen Question(s)	Incident Question(s)	Estimate
Rape (completed, attempted) Sexual coercion (completed, attempted) Sexual contact with force or without force (completed, attempted) Threats of each of the above	Two stages: (1) behaviorally specific screen questions; (2) detailed incident report (Since school began in fall 1996) Hierarchical scoring procedure used in incident report to classify type of crime, if any, that occurred	Multiple behaviorally specific screen questions (7, 8, 9, 10, 11, 12, 13, 14, 15, 16, 17, 18, 19, 20, 21, 22)	Responses to multiple items about type of completed, attempted, and threatened acts, penetration and physical contact; physical force used or threatened with physical force	Percentage Estimated counts

TABLE D-8 National College Women Sexual Victimization Study: Concept and Definitions

Concept	Definitions
Completed Rape	**Unwanted completed penetration** by physical force or the threat of physical force. Penetration includes penile-vaginal, mouth on your genitals, mouth on someone else's genitals, penile-anal, digital-vaginal, digital-anal, object-vaginal, and object-anal.
Attempted Rape	**Unwanted attempted penetration** by force or the threat of force.
Threat of Rape	**Threat of unwanted penetration** with force and threat of force.

NOTE: For other types of sexual violence: see https://www.ncjrs.gov/pdffiles1/nij/182369.pdf (pp. 6-8) [September 2013].
SOURCE: Data from Fisher et al. (2000).

SEXUAL VICTIMIZATION SCREEN QUESTIONS

Women may experience a wide range of unwanted sexual experiences in college. Women do not always report unwanted sexual experiences to the police or discuss them with family or friends. The person making the advances is not always a stranger, but can be a friend, boyfriend, fellow student, professor, teaching assistant, supervisor, co-worker, somebody you met off campus, or even a family member. The experience could occur anywhere: on- or off-campus, in your residence, in your place of employment, or in a public place. You could be awake, or you could be asleep, unconscious, drunk, or otherwise incapacitated. Please keep this in mind as you answer the questions.

Now, I'm going to ask you about different types of unwanted sexual experiences you may have experienced since school began in the Fall 1996. Because of the nature of unwanted sexual experience, the language may seem graphic to you. However, this is the only way to assess accurately whether or not the women in this study have had such experiences. You only have to answer "yes" or "no".

7. Since school began in the Fall 1996, has anyone **made** you have **sexual intercourse** by using **force or threatening to harm** you or someone close to you? Just so there is no mistake, by intercourse I mean putting a penis in your vagina.

Yes..............1
No....................2
Not sure.....3
Refused........4

8. Since school began in the Fall 1996, has anyone **made** you have **oral sex by force or threat of harm?** By oral sex, I mean did someone's mouth or tongue make contact with your vagina or anus or did your mouth or tongue make contact with someone else's genitals or anus.

Yes............1
No..................2
Not sure...3
Refused......4

9. Since school began in the Fall 1996, has anyone **made** you have **anal sex by force or threat of harm**? By anal sex, I mean putting a penis in your anus or rectum.

Yes............1
No.................2
Not sure...3
Refused......4

Data from Fisher et al. (2000).

10.Since school began in the Fall 1996, has anyone ever used **force or threat of harm to sexually penetrate you with a foreign object?** By this, I mean for example, placing a bottle or finger in your vagina or anus.

Yes............1
No................2
Not sure...3
Refused......4

CONDITIONAL: IF ANY "YES' IN Q7-10 CONTINUE, ELSE SKIP TO Q 12.

DUMMY: VAGINAL SEX
ORAL SEX
ANAL SEX
PENETRATION WITH OBJECTS

11. How many different incidents of forced (DUMMY) have happened to you since school began in FALL 1996?

__________NUMBER OF INCIDENTS

REFUSED........97
DK....................98

12.Since school began in Fall 1996, has anyone **attempted but not succeeded** in making you take part in any of the unwanted sexual experiences that I have just asked you about? *This would include threats that were not followed through.* For example, did anyone threaten or try but not succeed to have vaginal, oral, or anal sex with you or try unsuccessfully to penetrate your vagina or anus with a foreign object or finger?

Yes...............1
No...................2
Not sure.....3
Refused.......4

CONDITIONAL: IF Q12 EQ 1 CONTINUE, ELSE SKIP TO Q14

13. How many different incidents of unsuccessful attempts *or threats* of forced sex have happened to you since school began in FALL 1996?

__________NUMBER OF INCIDENTS
REFUSED......97
DK..................98

Data from Fisher et al. (2000).

14. Not counting the types of sexual contact already mentioned, have you experienced **any unwanted or uninvited touching of a sexual nature** since school began in the Fall 1996? This includes forced kissing, touching of private parts, grabbing, and fondling, even it is over your clothes. Remember this could include anyone from strangers to people you know well. Have any incidents of **unwanted or uninvited touching of a sexual nature** happened to you since school began in the Fall 1996?

Yes............1
No.................2
Not sure...3
Refused.......4

CONDITIONAL: IF Q14 EQ 1 CONTINUE, ELSE SKIP TO Q16

15. How many different incidents of unwanted or uninvited touching of a sexual nature have happened to you since school began in FALL 1996?

__________NUMBER OF INCIDENTS
REFUSED........97
DK....................98

16.Since school began in Fall 1996, has anyone **attempted** *or threatened* **but not succeeded** in **unwanted or uninvited touching of a sexual nature**?

Yes...............1
No...................2
Not sure.....3
Refused........4

CONDITIONAL: IF Q16 EQ 1 CONTINUE, ELSE SKIP TO Q18

17. How many different incidents of unwanted or uninvited attempts *or threats* at touching of a sexual nature have happened to you since school began in FALL 1996?

__________NUMBER OF INCIDENTS
REFUSED........97
DK....................98

Data from Fisher et al. (2000).

18. I have been asking you about unwanted sexual contact that involved force or threats of force against you or someone else. Sometimes unwanted sexual contact may be attempted using threats of nonphysical punishment, promises of rewards if you complied sexually, or simply continual verbal pressure.

Since school began in Fall 1996, has anyone made or tried to make you have sexual intercourse or sexual contact when you did not want to by making **threats of non-physical punishment** such as lowering a grade, being demoted or fired from a job, damaging your reputation or being excluded from a group for failure to comply with requests for any type of sexual activity.

Yes..............1
No...................2
Not sure.....3
Refused........4

19. Since school began in the Fall 1996, has anyone made or tried to make you have sexual intercourse or sexual contact when you did not want to by **making promises of rewards** such as raising a grade, being hired or promoted, being given a ride or class notes, or getting help with coarse work from a fellow student if you complied sexually.

Yes..............1
No...................2
Not sure.....3
Refused........4

20. Since school began in the Fall 1996, has anyone made or tried to make you have sexual intercourse or sexual contact when you did not want to by simply being **overwhelmed by someone's continual pestering and verbal pressure?**

Yes..............1
No...................2
Not sure.....3
Refused........4

CONDITIONAL: IF YES IN Q18, Q19 or Q20 ASK Q21, ELSE SKIP TO Q22
21. How many different incidents of non-physical threats, rewards or continual verbal pressure to make you have sexual intercourse or contact have *happened to you* since school began in FALL 1996?
__________NUMBER OF INCIDENTS
REFUSED..................97
DK...............................98

Data from Fisher et al. (2000).

22. Not counting any incidents we have already discussed, have you experienced any other type of unwanted or uninvited sexual contact since school began in the Fall? Remember, this could include sexual experiences that may or may not have been reported to the police or other officials, which were with strangers or people you know, in variety of locations both on- and off-campus, and while you were awake, or when you were asleep, drunk, or otherwise incapacitated.

Yes..............1
No...................2
Not sure.....3
Refused........4

CONDITIONAL: IF YES IN Q22 ASK Q23, ELSE SKIP TO Q24
23. How many different incidents of these other types of unwanted or uninvited sexual contact have happened to you since school began in Fall 1996?

__________NUMBER OF INCIDENTS
REFUSED........97
DK....................98

SEXUAL INCIDENT COUNTER:

A. NUMBER OF FORCED SEXUAL PENETRATIONS FROM Q11
B. NUMBER OF ATTEMPTED/THREATENED FORCED SEXUAL PENETRATIONS FROM Q13
C. NUMBER OF SEXUAL TOUCHINGS OR ASSAULTS FROM Q15
D. NUMBER OF ATTEMPTED/THREATENED SEXUAL TOUCHING OR ASSAULT FROM Q17
E. NUMBER OF SEXUAL COERCION OR PRESSURE FROM Q21
F. NUMBER OF OTHER UNWANTED SEXUAL CONTACTS FROM Q23

COMPUTE TOTAL INCIDENTS A-F. IF COUNTER EQ 0, SKIP TO Q24
IF SEXUAL INCIDENT COUNTER IS GREATER THAN ZERO, GO TO INCIDENT REPORT LOOP.
INCIDENT REPORT WILL BE COMPLETED FOR EACH REPORTED INCIDENT BY CATEGORY BEGINNING WITH THE MOST RECENT INCIDENT IN THAT CATEGORY. THERE IS A MAXIMUM OF FIVE LOOPS PER CATEGORY.

Data from Fisher et al. (2000).

NATIONAL INTIMATE PARTNER AND SEXUAL VIOLENCE SURVEY

The National Intimate Partner and Sexual Violence Survey is supported by the National Institute of Justice and the Centers for Disease Control and Prevention. It was first fielded in 2010; the two agencies plan to conduct it on a regular basis (see Chapter 5). Table D-9 presents details about its design and estimates and Table D-10 shows the survey's concept and its definition. They are followed by the victimization questions used in the survey.

TABLE D-9 The National Intimate Partner and Sexual Violence Survey: Design and Estimates

Design		Data Collection	
Research Design	Population/ Sampling Design, Sample Size, Age Range	Mode of Administration	Framing of Survey Context
Cross-section	Adults in United States (ages 18 and older) A dual-frame sampling strategy including both landline and cell phones in 50 states National random digit dialing *n* = 18,049 Male population: 77,814,000 Female population: 139,808,000	Computer-assisted telephone interviewing	Health

SOURCE: Data from Black et al. (2011).

Form of Sexual Victimizations	Operationalization			
Type	Measurement Process (reference frame)	Screen Question(s)	Incident Question(s)	Estimate
Rape; completed, forced penetration, attempted penetration, alcohol- or drug-facilitated completed penetration Sexual coercion Unwanted sexual contact Noncontact unwanted sexual experiences	Single stage Behaviorally specific (Lifetime prevalence of rape) (Prevalence of rape in the 12 months prior to taking the survey)	21 single items	Single-stage classification with no separate incident report	Percentages Frequencies

TABLE D-10 National Intimate Partner and Sexual Violence Survey (NISVS): Concept and Definitions

Concept	Definitions
Rape	**Rape** is defined as any completed or attempted unwanted vaginal (for women), oral, or anal penetration through the use of physical force (such as being pinned or held down, or by the use of violence) or threats to physically harm and includes times when the victim was drunk, high, drugged, or passed out and unable to consent. Rape is separated into three types: completed forced penetration, attempted forced penetration, and completed alcohol- or drug-facilitated penetration. —Among women, rape includes vaginal, oral, or anal penetration by a male using his penis. It also includes vaginal or anal penetration by a male or female using his or her fingers or an object. —Among men, rape includes oral or anal penetration by a male using his penis. It also includes anal penetration by a male or female using his or her fingers or an object.
Sexual coercion	**Sexual coercion** is defined as unwanted sexual penetration that occurs after a person is pressured in a nonphysical way. In NISVS, sexual coercion refers to unwanted vaginal, oral, or anal sex after being pressured in ways that included being worn down by someone who repeatedly asked for sex or showed they were unhappy; feeling pressured by being lied to, being told promises that were untrue, or having someone threaten to end a relationship or spread rumors; and sexual pressure due to someone using his or her influence or authority.
Unwanted sexual contact	**Unwanted sexual contact** is defined as unwanted sexual experiences involving touch but not sexual penetration, such as being kissed in a sexual way or having sexual body parts fondled or grabbed.
Noncontact unwanted sexual experiences	**Noncontact unwanted sexual experiences** are those unwanted experiences that do not involve any touching or penetration, including someone exposing his or her sexual body parts, flashing, or masturbating in front of the victim, someone making a victim show his or her body parts, someone making a victim look at or participate in sexual photos or movies, or someone harassing the victim in a public place in a way that made the victim feel unsafe.

SOURCE: Data from Black et al. (2011).

Victimization Questions

Sexual Violence	
How many people have ever…	• exposed their sexual body parts to you, flashed you, or masturbated in front of you? • made you show your sexual body parts to them? Remember, we are only asking about things that you didn't want to happen. • made you look at or participate in sexual photos or movies?
How many people have ever…	• harassed you while you were in a public place in a way that made you feel unsafe? • kissed you in a sexual way? Remember, we are only asking about things that you didn't want to happen. • fondled or grabbed your sexual body parts?
When you were drunk, high, drugged, or passed out and unable to consent, how many people ever…	• had vaginal sex with you? By vaginal sex, we mean that {if female: a man or boy put his penis in your vagina} {if male: a woman or girl made you put your penis in her vagina}? • {if male} made you perform anal sex, meaning that they made you put your penis into their anus? • made you receive anal sex, meaning they put their penis into your anus? • made you perform oral sex, meaning that they put their penis in your mouth or made you penetrate their vagina or anus with your mouth? • made you receive oral sex, meaning that they put their mouth on your {if male: penis} {if female: vagina} or anus?
How many people have ever used physical force or threats to physically harm you to make you…	• have vaginal sex? • {if male} perform anal sex? • receive anal sex? • make you perform oral sex? • make you receive oral sex? • put their fingers or an object in your {if female: vagina or} anus?
-How many people have ever used physical force or threats of physical harm to…	• {if male} try to make you have vaginal sex with them, but sex did not happen? • try to have {if female: vaginal} oral, or anal sex with you, but sex did not happen?
-How many people have you had vaginal, oral, or anal sex with after they pressured you by…	• doing things like telling you lies, making promises about the future they knew were untrue, threatening to end your relationship, or threatening to spread rumors about you? • wearing you down by repeatedly asking for sex, or showing they were unhappy? • using their authority over you, for example, your boss or your teacher?

SOURCE: Data from Black et al. (2011).

NATIONAL SURVEY OF INTER-GENDER RELATIONSHIPS

The National Survey of Inter-Gender Relationships was supported by the Antisocial and Criminal Behavior Branch of the National Institute of Mental Health. It was conducted in 1984-1984 as a self-report questionnaire to a national sample of students enrolled in 32 institutions of higher education across the United States. Table D-11 presents details about its design and estimates. It is followed by the detailed questionnaires used in the survey.

TABLE D-11 The National Survey of Inter-Gender Relationships: Design and Estimates

Design		Data Collection	
Research Design	Population/ Sampling Design, Sample Size, Age Range	Mode of Administration	Framing of Survey Context
Cross-section	Probability sample of 4-year colleges and universities Two stages: (1) selection of institution; (2) selection of classes within institutions n = 6,159 students Female: 3,187 Male: 2,972 Mean age of male: 21.0 Mean age of female: 21.4	Self-report questionnaire	Sexual experience survey

SOURCE: Data from Koss, Gidycz, and Wisniewski (1987).

Form of Sexual Victimization	Operationalization			
Type	Measurement Process (time frame)	Screen Question(s)	Incident Question(s)	Estimate
Sexual contact (by verbal coercion, misuse of authority, threat or force) Attempted intercourse (by force, alcohol, or drugs) Sexual coercion (intercourse by verbal coercion, misuse of authority, alcohol or drugs, threat or force) Rape (oral or anal penetration by threat or force)	One stage: behaviorally specific (From age 14) (Last school year) Hierarchical scoring procedure (In the past 12 months) (Lifetime prevalence of rape)	Multiple single items Sexual contact (items 1, 2, 3) Attempted rape (items 4, 5) Sexual coercion (items 6, 7) Rape (items 8, 9, 10)	Single-stage classification with no separate incident report	Percentage Frequencies Mean (SD)

SEXUAL BEHAVIORS

1. Have you given in to sex play (fondling, kissing, or petting, but not intercourse) when you didn't want to because you were overwhelmed by a man's continual arguments and pressure?
2. Have you had sex play (fondling, kissing, or petting, but not intercourse) when you didn't want to because a man used his position of authority (boss, teacher, camp counselor, supervisor) to make you?
3. Have you had sex play (fondling, kissing, or petting, but not intercourse) when you didn't want to because a man threatened or used some degree of physical force (twisting your arm, holding you down, etc.) to make you?
4. Have you had a man attempt sexual intercourse (get on top of you, attempt to insert his penis) when you didn't want to by threatening or using some degree of force (twisting your arm, holding you down, etc.), but intercourse did not occur?
5. Have you had a man attempt sexual intercourse (get on top of you, attempt to insert his penis) when you didn't want to by giving you alcohol or drugs, but intercourse did not occur?
6. Have you given in to sexual intercourse when you didn't want to because you were overwhelmed by a man's continual agruments and pressure?
7. Have you had sexual intercourse when you didn't want to because a man used his position of authority (boss, teacher, camp counselor, supervisor) to make you?
8. Have you had sexual intercourse when you didn't want to because a man gave you alcohol or drugs?
9. Have you had sexual intercourse when you didn't want to because a man threatened or used some degree of physical force (twisting your arm, holding you down, etc.) to make you?
10. Have you had sex acts (anal or oral intercourse or penetration by objects other than the penis) when you didn't want to because a man threatened or used some degree of physical force (twisting your arm, holding you down, etc.) to make you?

SOURCE: Data from Koss, Gidycz, and Wisniewski (1987).

NATIONAL WOMEN'S STUDY—REPLICATE

The National Women's Study—Replicate was conducted in 2006 with support from the National Institute of Justice. The results were published in Kilpatrick et al. (2007). Table D-12 presents details about its design and estimates and Table D-13 shows the survey's concept and its description. They are followed by the list of screening questions used in the survey.

TABLE D-12 National Women's Study-Replication: Design and Estimates

Design		Data Collection	
Research Design	Population/ Sampling Design, Sample Size, Age Range	Mode of Administration	Framing of Survey Context
Cross-section	Two groups: (1) General population, adult-household residing women living in United States/random digit dialing methods n = 3,001 (ages 18 to 86 years; with younger women oversampled, mean = 46.6) (2) Adult women enrolled in 4-year institutions of higher education in United States/classification of sample by nine regions and sample released to be dialed in proportion to the national census representation of college women n = 2,000; n = 253 different schools (ages 18 to 67 years, mean = 20.1)	Telephone survey—computer-assisted telephone interviewing	Unwanted sexual advances

SOURCE: Data from Kilpatrick et al. (2007).

Form of Sexual Victimization	Operationalization			
Type(s)/ Definitions	Measurement Process (reference frame)	Screen Question(s)	Incident Question(s)	Estimate
Drug- and alcohol-facilitated rape Incapacitated rape Forcible rape	Single stage Behaviorally specific (Lifetime and past year prevalence of rape)	See following pages	For one reported incident or if more than one, most recent	Lifetime and 2005 annual prevalence estimates for U.S. population of women, and separately for college women Estimates for total rape, forcible rape, drug- and alcohol-facilitated rape, and incapacitated rape

TABLE D-13 National Women's Study-Replication: Concept and Description

Concept	Description*
Drug- and alcohol-facilitated rape	**Drug- and alcohol-facilitated rape (DAFR)** is when the perpetrator deliberately gives the victim drugs without her permission or tries to get her drunk, and then commits an unwanted sexual act against her involving oral, anal, or vaginal penetration. The victim is passed out or awake but too drunk or high to know what she is doing or to control her behavior.
Incapacitated rape	**Incapacitated rape (IR)** is unwanted sexual act involving oral, anal, or vaginal penetration that occurs after the victim voluntarily uses drugs or alcohol. The victim is passed out or awake but too drunk or high to know what she is doing or to control her behavior.
Forcible rape	**Forcible rape (FR)** is unwanted sexual act involving oral, anal or vaginal penetration. The victim also experiences force, threat of force, or sustains an injury during the assault. In cases where FR includes elements of DAFR, incident categorized as DAFR.

*By definition, DAFR and IR are mutually exclusive.
SOURCE: Data from Kilpatrick et al. (2007).

LIST OF RAPE SCREENING QUESTIONS USED IN THE INTERVIEW

Our interviewers read, "Many women tell us they have experienced unwanted sexual advances at some point during their lives. Women do not always report such experiences to police or discuss them with family or friends. Such experiences can happen anytime in a woman's life—even as a child. The person making these unwanted advances can be friends, boyfriends, co-workers, teaching assistants, supervisors, family members, strangers, or someone they just met. The person making the unwanted sexual advances can be male or female. . . . Regardless of how long ago it happened or who made the unwanted sexual advances:

1. Has a man or boy ever made you have sex by using force or threatening to harm you or someone close to you? Just so there is no mistake, by having sex, we mean putting a penis in your vagina.
2. Has anyone, male or female, ever made you have oral sex by force or threatening to harm you? So there is no mistake, by oral sex, we mean that a man or boy put his penis in your mouth or someone penetrated your vagina or anus with their mouth or tongue?
3. Has anyone ever made you have anal sex by force or threatening to harm you? By anal sex, we mean putting their penis in your anus or rectum.
4. Has anyone ever put fingers or objects in your vagina or anus against your will by using force or threatening to harm you?

Some women tell us they have had sex when they didn't want to because they were very high, intoxicated, or even passed out because of alcohol or drugs. We would like to ask you about these types of experiences you might have had. Again, we are interested in these experiences regardless of how long ago it happened, who did it, or whether or not it was reported to police.

1. Has anyone ever had sex with you when you didn't want to after you drank so much alcohol that you were very high, drunk, or passed out? By having sex, we mean that a man or boy put his penis in your vagina, your anus, or your mouth?
2. Has anyone ever had sex with you when you didn't want to after they gave you, or you had taken enough drugs to make you very high, intoxicated, or passed out? By having sex we mean that a man or boy put his penis in your vagina, your anus, or your mouth?

Sample of follow-up questions asked upon endorsement of one or more screeners:

- Were you physically forced to engage in these acts?
- Did the person or persons who did this to you threaten to hurt you or someone else if you did not do what they wanted?
- Had you ever seen the person who did this to you before?
- Did you know the person fairly well or not?
- Had you consumed any drugs or alcohol at the time of the incident(s)?
- When this happened, did the incident involve only alcohol use on your part, only drug use on your part, or some use of both alcohol and drugs?
- When this happened, did you drink the alcohol because you wanted to, did the person(s) who had sex with you deliberately try to get you drunk, or both?
- When this happened, did you take the drugs because you wanted to, did the person(s) who had sex with you deliberately give you drugs without your permission, or both?
- When this incident happened were you passed out from drinking or taking drugs?
- When this incident happened were you awake but too drunk or high to know what you were doing or control your behavior?
- Did you suffer serious physical injuries, minor injuries, or no physical injuries as a result of the incident?
- Did this incident involve oral penetration, anal penetration, or vaginal penetration?

For full questionnaire, see http://www.icpsr.umich.edu/icpsrweb/NACJD/studies/20626.

SOURCE: Data from Kilpatrick et al. (2007).

CAMPUS SEXUAL ASSAULT STUDY

The Campus Sexual Assault Study was conducted in 2006 with support from the National Institute of Justice. The results were published in Krebs et al. (2007). Table D-14 presents details about its design and estimates and Table D-15 shows the survey's concept and its description. These data are followed by the questions used in the survey and the classification procedure.

TABLE D-14 Campus Sexual Assault Study: Design and Estimates

Design		Data Collection	
Research Design	Population/ Sampling Design, Sample Size, Age Range	Mode of Administration	Framing of Survey Context
Cross-section	Undergraduate students enrolled at least three quarters time during 2005-2006 academic year at two large public universities Probability sample Total n = 5,446 Males: n = 1,375 Females: n = 5,446 (ages 18-25)	Web-based survey	Nonconsensual or unwanted sexual contact experience

SOURCE: Data from Krebs et al. (2007).

Form of Sexual Victimization	Operationalization			
Type(s)	Measurement Process (reference frame)	Screen Question(s)	Incident Question(s)	Estimate
Physically forced sexual assault Incapacitated sexual assault	One stage: behaviorally specific items (Before you began college) (Since you began college)	Responses from multiple items used	Single-stage classification with no separate incident report	Percentages Frequencies

TABLE D-15 Campus Sexual Assault Study: Concept and Descriptions

Concept	Description(s)
Physically forced sexual assault	**Physically forced sexual assault** includes assaults occurring when the victim was forced or threatened with force into sexual contact. Forced touching of a sexual nature (forced kissing, touching of private parts, grabbing, fondling, rubbing up against you in a sexual way, even if it is over your clothes). Oral sex (someone's mouth or tongue making contact with your genitals or your mouth or tongue making contact with someone else's genitals). Sexual intercourse (someone's penis being put in your vagina). Anal sex (someone's penis being put in your anus). Sexual penetration with a finger or object (someone putting their finger or an object like a bottle or a candle in your vagina or anus).
Incapacitated sexual assault	Assaults occurring when a victim is unable to provide consent or stop what is happening because she is passed out, drugged, drunk, incapacitated, or asleep. **Alcohol or other drug–enabled sexual assault,** a subset of incapacitated assault, occurs when the victim is incapacitated after voluntarily consuming alcohol and/or drugs. **Drug-facilitated sexual assault,** another subset of incapacitated assault, occurs when the victim is incapacitated after being given a drug without her knowledge or consent.

SOURCE: Data from Krebs et al. (2007).

Part 1. Interview Questions Used in Sexual Assault Classification

This section of the interview asks about nonconsensual or unwanted sexual contact you may have experienced. When you are asked about whether something happened since you began college, please think about what has happened since you entered any college or university. The person with whom you had the unwanted sexual contact could have been a stranger or someone you know, such as a family member or someone you were dating or going out with.

These questions ask about five types of unwanted sexual contact:

- forced touching of a sexual nature (forced kissing, touching of private parts, grabbing, fondling, rubbing up against you in a sexual way, even if it is over your clothes)
- oral sex (someone's mouth or tongue making contact with your genitals or your mouth or tongue making contact with someone else's genitals)
- sexual intercourse (someone's penis being put in your vagina)
- anal sex (someone's penis being put in your anus)
- sexual penetration with a finger or object (someone putting their finger or an object like a bottle or a candle in your vagina or anus.

The questions below ask about unwanted sexual contact that involved force or threats of force against you. Force could include someone holding you down with his or her body weight, pinning your arms, hitting or kicking you, or using or threatening to use a weapon against you.

	Before you began college	Since you began college
Has anyone had sexual contact with you by using physical force or threatening to physically harm you?	○ Yes ○ No [V9]*	○ Yes ○ No [V3]
Has anyone attempted but not succeeded in having sexual contact with you by using or threatening to use physical force against you?	○ Yes ○ No [V10]	○ Yes ○ No [V4]

SOURCE: Data from Krebs et al. (2007).

The next set of questions ask about your experiences with unwanted sexual contact while you were unable to provide consent or stop what was happening because you were passed out, drugged, drunk, incapacitated, or asleep. These situations might include times that you voluntarily consumed alcohol or drugs and times that you were given drugs without your knowledge or consent.

	Before you began college	**Since you began college**
Has someone had sexual contact with you when you were unable to provide consent or stop what was happening because you were passed out, drugged, drunk, incapacitated, or asleep? This question asks about incidents that you are <u>certain</u> happened.	○ Yes ○ No [V11]	○ Yes ○ No [V5]
Have you <u>suspected</u> that someone has had sexual contact with you when you were unable to provide consent or stop what was happening because you were passed out, drugged, drunk, incapacitated, or asleep? This question asks about events that you think (but are not certain) happened.	○ Yes ○ No [V12]	○ Yes ○ No [V6]

[If V3=yes] Earlier you indicated that since you began college, someone has had sexual contact with you by using physical force or threatening to physically harm you. The questions below ask about that experience.

V3b. When the person had sexual contact with you by using or threatening you with physical force, which of the following happened? Please check all that apply.

- □ Forced touching of a sexual nature
- □ Oral sex
- □ Sexual intercourse
- □ Anal sex
- □ Sexual Penetration with a finger or object

SOURCE: Data from Krebs et al. (2007).

[If V5=yes] Earlier you indicated that since you began college, someone has had sexual contact with you when you were unable to provide consent or stop what was happening because you were passed out, drugged, drunk, incapacitated, or asleep. The questions below ask about that experience.

V5b. [if V5=yes] When the person had sexual contact with you when you were unable to provide consent or stop what was happening because you were passed out, drugged, drunk, incapacitated, or asleep, which of the following happened? Please check all that apply.

- ☐ Forced touching of a sexual nature
- ☐ Oral sex
- ☐ Sexual intercourse
- ☐ Anal sex
- ☐ Sexual Penetration with a finger or object
- ☐ Don't Know

[If V5=yes] The next questions ask more about the time <u>since you entered college</u> that someone had sexual contact with you when you were unable to provide consent or stop what was happening because you were passed out, drugged, drunk, incapacitated, or asleep.

C27. Just prior to (the incident/any of the incidents), had you been drinking alcohol? Keep in mind that you are in no way responsible for the assault that occurred, even if you had been drinking.

- ○ Yes
- ○ No

C27a. [if C27=yes or 99] Were you drunk?

- ○ Yes
- ○ No

C28. Just prior to (the incident/any of the incidents), had you voluntarily been taking or using any drugs other than alcohol?

- ○ Yes
- ○ No

C29. Just prior to (the incident/any of the incidents), had you been given a drug without your knowledge or consent? (yes, no, don't know)

- ○ Yes
- ○ No
- ○ Don't Know

SOURCE: Data from Krebs et al. (2007).

Appendix E

Statistical Rationale Behind Some Initial Findings on the Relative Statistical Plausibility of a Multiple-Frame Approach to Estimating the Victimization Rate of Rape and Sexual Assault

William D. Kalsbeek[1]

This paper expands the discussion in Chapter 10 on the use of a multiple-frame approach to estimating the incidence of rape and sexual assault in household surveys of the Bureau of Justice Statistics. It explores the statistical rationale behind some initial findings on the relative statistical plausibility of a multiple-frame approach.[2]

BACKGROUND AND ASSUMPTIONS

1. The primary analysis objective is to estimate the proportion (P) of persons in the target population who have been a victim of a rape or sexual assault (RSA) in some calendar year.
2. The following two overlapping frames are involved in defining a dual-frame (**DF**) sample design that might be used to estimate P: (1) an administrative frame consisting of persons seen/treated/processed for their RSA during the same calendar year and (2) a standard area household frame of the residential population of the kind used for the NCVS.

[1]Kalsbeek is a professor in the Department of Biostatistics at the University of North Carolina. He served as cochair of this panel.

[2]A presentation on the statistical issues in this appendix was presented at the Joint Statistical Meetings in Montreal in August 2013 (Kalsbeek, Spencer, and House, 2013), available http://www.amstat.org/meetings/jsm/2013/onlineprogram/AbstractDetails.cfm?abstractid=309226 [December 2013].

3. The administrative frame is a subset of the area household frame, and thus the two frames overlap. However, one can define two nonoverlapping strata by considering those in the administrative frame to be one stratum and all members of the area household frame not included in the administrative frame to be the second stratum, implying that a sample for the second stratum selected from the area household frame would need to be screened to excluded members of the administrative frame. Formation of these two strata is the simplest frame construction arrangement for a dual-frame design and comparable to the frame structure of telephone sampling of landline and cell-only households (Hartley, 1962; Lohr, 2011).
4. The administrative frame might be chosen from any of the following sets of people who: (1) filed a crime complaint with the police or some other law enforcement agency, (2) were victims of RSA or aggravated assault when an accused perpetrator is charged with a crime and tried in the criminal justice system, (3) were treated for assault-related health consequences by a hospital emergency department, (4) were clients of victim support services (e.g., rape crisis center, domestic violence shelters, etc.), (5) were registered residents of Indian reservations, (6) were treated at Indian Health Services facilities, or (7) were patients of outpatient mental health clinics.
5. A simple form of sampling (i.e., simple random sampling with replacement, SRSWR) is applied separately to the administrative and the nonadministrative household strata.
6. The dual-frame sample design is seen as an alternative to a single-frame (**SF**) design but uses a standard area household frame as currently used in the NCVS. While more complex forms of stratified cluster sampling would be used with **DF** and **SF** designs, one assumes SRSWR sampling is applied to each frame, with the presumption that effects of greater sampling complexity would cancel, thus sustaining a comparison between the two design alternatives.

DETERMINING THE MOST COST-EFFICIENT SAMPLE ALLOCATION AMONG STRATA IN THE DUAL-FRAME DESIGN

One can consider the simplest case of multiframe sample design in which the set of population members comprising two overlapping frames is divided into two nonoverlapping sampling strata, as for instance with cell and landline frames in telephone sampling (Hartley, 1962; Lohr, 2011). In the situation described above, we have two nonoverlapping sampling strata formed by the members of: (1) the administrative frame (A), and (2) the nonadministrative household frame (HH) consisting of those members of the *HH* frame who are not members of the administrative frame. Under this scenario one can observe the precision of a dual-frame estimator of the prevalence of rape and sexual assault on the basis of well-known properties of the analysis from a stratified sample.

For stratified SRSWR, the variance of the estimator, $p_W = \sum_{h=1}^{H} W_h p_h$, of P for the general case of selecting a sample of size n from H strata is

$$V(p_W) = \sum_{h=1}^{H} W_h^2 V(p_h) = \sum_{h=1}^{H} W_h^2 \frac{P_h(1-P_h)}{n_h},$$

where for the *h-th* stratum: $W_h = N_h/N$ is the proportion of the population, P_h is the proportion of victims of RSA among all N_h population members, and p_h is the proportion of RSA victims among the n_h sample members. If one defines C_h, the average cost of adding another survey respondent in the *h-th* stratum, then we can use the simple linear variable cost model, $C^* = \sum_{h=1}^{H} C_h n_h$ and the Cauchy-Schwartz inequality to establish the sample allocation that minimizes $\left[\{V(p_W)\} x \{C^*\}\right]$. The most cost-efficient sample allocation to the *h-th* stratum is thereby

$$n_h^{(C-E)} = K \frac{W_h \sqrt{P_h(1-P_h)}}{\sqrt{C_h}}, \quad [1]$$

where $K = \dfrac{C^*}{\sum_{h=1}^{H} W_h \sqrt{P_h(1-P_h)}\sqrt{C_h}}$.

Applying the general result from Eq. [1] to the two-stratum setting of the dual frame,

$$n_A^{(C-E)} = K \frac{W_A \sqrt{P_A(1-P_A)}}{\sqrt{C_A}} \quad [2]$$

for the administrative stratum, and

$$n_{HH}^{(C-E)} = K \frac{W_{HH} \sqrt{P_{HH}(1-P_{HH})}}{\sqrt{C_{HH}}} \quad [3]$$

for the household stratum, where

$$K = \frac{C^*}{W_A \sqrt{P_A(1-P_A)}\sqrt{C_A} + W_{HH} \sqrt{P_{HH}(1-P_{HH})}\sqrt{C_{HH}}}.$$

VARIANCE OF A DUAL-FRAME ESTIMATE BASED ON THE MOST COST-EFFICIENT ALLOCATION

The variance of p_W for the stratified SRSWR with the most cost-efficient sample allocation (i.e., the $n_h^{(C\text{–}E)}$) for the case of H strata can be shown to be

$$V_{DF}^{(C-E)}(p_W) = \frac{\left[\sum_{h=1}^{H} W_h \sqrt{P_h(1-P_h)} / \sqrt{C_h}\right]}{n / \left[\sum_{h=1}^{H} W_h \sqrt{P_h(1-P_h)}\sqrt{C_h}\right]}.$$

For the two-stratum case,

$$V_{DF}^{(C-E)}(p_W) = \frac{\left[W_A \sqrt{P_A(1-P_A)} / \sqrt{C_A} + W_{HH}\sqrt{P_{HH}(1-P_{HH})} / \sqrt{C_{HH}}\right]}{n / \left[W_A \sqrt{P_A(1-P_A)}\sqrt{C_A} + W_{HH}\sqrt{P_{HH}(1-P_{HH})}\sqrt{C_{HH}}\right]}. \quad [4]$$

Dual-Frame vs. Single-Frame *HH* Area Household Frame Design

A cost-equivalent comparison of the dual-frame (**DF**) estimator with a single-frame (**SF**) estimator with a sample of size $n_{SF} = C^*/C_{HH}$ when the total variable cost of data collection for the **SF** design is C^*. For design comparability one assumes SRSWR sampling from the household frame in which case the variance of the **SF** estimator (p_{HH}) of P will be simply

$$V_{SF}(p_{HH}) = \frac{P(1-P)}{n_{SF}}. \quad [5]$$

The variances of estimates of P by the **DF** and **SF** designs can be compared using the ratio

$$R_V = \frac{V_{DF}^{(C-E)}(p_W)}{V_{SF}(p_{HH})}. \quad [6]$$

Other Comparison Indicators

1. *Ratio of Average Unit Costs for the Two Dual-Frame Strata*—This ratio depicts the ratio of the average cost of adding another respondent to the administrative stratum compared to the comparable average cost for the nonadministrative household stratum. This indicator is computed as

$$\theta = \frac{C_A}{C_{HH}}. \quad [7]$$

2. *Ratio of Stratum RSA Rates for the Dual-Frame Design*—Compared to an unstratified SRSWR design, Cochran (1977, Section 5.6) notes that when stratum unit costs are equal the relative effectiveness of the most cost-efficient stratum allocation for a stratified SRSWR depends on the extent of stratum differences in (i) P_h and (ii) the standard error of the RSA status (i.e., $\sigma_{\wp} = \sqrt{P_h(1-P_h)}$). Differences in (ii) are especially pronounced for extremely small (or large) values of P_h, as is the case here with P being about *0.001* for the rate of RSA prevalence, and thus implying that $P_A >> P_{HH}$. The indicator used to measure the relative sizes of P_A and P_{HH} is

$$\beta = \frac{P_A}{P_{HH}}. \quad [8]$$

3. *Extent of Oversampling Members of the Administrative Frame in the Dual-Frame Design*—This is a descriptive indicator of the relatively greater sampling intensity in the administrative stratum compared to the household stratum in the DF design. The indicator is computed as

$$\phi = \frac{f_A^{(C-E)}}{f_H^{(C-E)}} = \frac{n_A^{(C-E)} / N_A}{n_{HH}^{(C-E)} / N_{HH}}. \quad [9]$$

4. *Percentage of Dual-Frame Sample from Administrative Stratum*—Indicates how much of the total dual-frame sample (n_{DF}) comes from the administrative frame. The indicator is computed as

$$Admin\ Sample\ \% = 100\, x\ \frac{n_A^{(C-E)}}{n_{DF}}. \quad [10]$$

5. *Relative Size of the Dual-Frame Sample Compared to the Single-Frame Sample*— Indicates the comparative sizes of the total sample sizes for the **DF** design (n_{DF}) vs. the **SF** design (n_{DF}). The indicator is computed as

$$Relative\ Overall\ Sample\ Size = n_{DF} / n_{SF}. \quad [11]$$

6. *Relative Standard Error of the Estimate for the Dual-Frame Design*—Relative measure of the precision of the dual-frame estimate with the most cost-efficient stratum allocation. The indicator is computed as

$$RSE_{DF}^{(C-E)}(p_W) = \frac{\sqrt{V_{DF}^{(C-E)}(p_W)}}{P}. \quad [12]$$

EXAMPLE 1: [$\theta = C_A/C_{HH} = 2$]

Suppose the following setting in which we are to compare the statistical quality of estimates from a **DF** design involving police records as the administrative source with comparable (and thus cost-equivalent) estimates from a household **SF** design as currently used in the NCVS. To determine the relative utility of **DF** and **SF** designs one might pose this question. How would the variance of a DF estimate of RSA prevalence ($V_{DF}^{(C-E)}(p_W)$) compare with the variance of a comparable **SF** estimate ($V_{SF}(p_{HH})$) obtained for the same cost?

To find an answer to this question within the context of the design assumptions, definitions, and theoretical findings described previously in this document, consider the following numerical values:

1. Police records are to be used to define an administrative stratum of crime victims, so specify the size of the administrative stratum as about N_A = *140,000* by extrapolating to the total U.S. population the 1997 Uniform Crime Reports partial national count of *96,122* assaults/attempts to commit rape as reported on p. 25 of *Crime in the United States 1997* (Federal Bureau of Investigation, 1997) at: http://www.fbi.gov/about-us/cjis/ucr/crime-in-the-u.s/1997/toc97.pdf.
2. From an August *BJS Selected Findings* report by CM Rennison (Bureau of Justice Statistics, 2002b) at: http://bjs.ojp.usdoj.gov/content/pub/pdf/rsarp00.pdf, the NCVS estimated average annual number of RSAs reported to police (1992-2000) was about *116,300*. Thus, the proportion of police records on assaults/attempts to commit rape that would turn out to be an RSA would be about P_A = *116,300/140,000 = 0.83*.[3]
3. Persons living at addresses define the household frame (as in the NCVS). According to Bureau of Justice Statistics (2008a) the total number of persons 12+ years of age is about N = *250,000,000* (in 2007), thus making the size of the household stratum $N_{HH} = N - N_A$ = *249,860,000*, and the proportion of the population in the administrative stratum will be about $W_A = 1 - W_{HH}$ = *140,000/250,000,000 = 0.00056*.
4. P = *0.001* based on figures from *Criminal Victimization, 2007* (Bureau of Justice Statistics, 2008a), which can be found at http://bjs.ojp.usdoj.gov/content/pub/pdf/cv07.pdf.
5. Based on a 2009 FCSM Research Conference paper presented

[3]If for confidentiality protection the types of crimes sampled through police records was broader, then P_A would be lower, and perhaps much lower, than this value.

by Michael R. Rand of BJS (Rand, 2009) in (see pages 9 and 16 of this paper) at http://www.fcsm.gov/09papers/Rand_X-B.doc, funds available to conduct the NCVS in FY2009 amounted to $C^* = \$26M$, and about 150,000 NCVS interviews were completed in 2008. These figures imply an average cost per completed interview of about $C_{HH} = \$26M/150000 = \$173$ for the household stratum.

Dual-Frame Design:

If the average per completed interview for the police records (administrative) stratum is two (2) times that of the household stratum (i.e., like the NCVS), then $\theta = C_A/C_M = 2$ and thus $C_A = \$346$.

First determine the RSA rate for the household stratum as $P_{HH} = \dfrac{P - W_A P_A}{1 - W_A} = 0.00053550$, which makes $P_A = 0.83$ larger than P_{HH} by a factor of about $\beta = \dfrac{P_A}{P_{HH}} \approx 1{,}550$. The standard deviations of the 0/1 RSA status indicator for the two strata thus differ by a factor of $\sigma_A / \sigma_{HH} = \sqrt{P_A(1-P_A)} / \sqrt{P_{HH}(1-P_{HH})} \approx 16.2$. Because of these substantial stratum differences in P_h and $\sigma_h = \sqrt{P_h(1-P_h)}$ one might expect from Eq. (5.37) in Cochran (1977) that a cost-efficient stratum allocation in this dual-frame context will produce substantially greater precision in estimates of P than a single-frame approach relying solely on household sampling. We will see this to be case below.

Using Equations [2] and [3] above, we find that the most cost-efficient allocation of the dual-frame sample given C^* for the police records stratum will be

$$n_A^{(C-E)} = \left[\frac{C^*}{W_A\sqrt{P_A(1-P_A)}\sqrt{C_A} + W_{HH}\sqrt{P_{HH}(1-P_{HH})}\sqrt{C_{HH}}}\right]\left[\frac{W_A\sqrt{P_A(1-P_A)}}{\sqrt{C_A}}\right] \approx 955$$

and for the household stratum,

$$n_{HH}^{(C-E)} = \left[\frac{C^*}{W_A\sqrt{P_A(1-P_A)}\sqrt{C_A} + W_{HH}\sqrt{P_{HH}(1-P_{HH})}\sqrt{C_{HH}}}\right]\left[\frac{W_{HH}\sqrt{P_{HH}(1-P_{HH})}}{\sqrt{C_{HH}}}\right] \approx 148{,}380.$$

Thus, the total sample size for the **DF** design in this case would be

149,334, of which *955* (or about 0.6%) would be from the police records stratum.

The variance of the weighted estimate of P from the **DF** design based on this most cost-efficient sample allocation between strata will be

$$V_{DF}^{(C-E)}(p_W) = \frac{\left[W_A\sqrt{P_A(1-P_A)}/\sqrt{C_A} + W_{HH}\sqrt{P_{HH}(1-P_{HH})}/\sqrt{C_{HH}}\right]}{n/\left[W_A\sqrt{P_A(1-P_A)}\sqrt{C_A} + W_{HH}\sqrt{P_{HH}(1-P_{HH})}\sqrt{C_{HH}}\right]} = 3.649\times 10^{-9}.$$

Cost-Equivalent Single-Frame Design:

Now turning our attention to the **SF** design, also with a budget of $C^* = \$26M$ and $C_{HH} = \$173$, the sample size we can afford for the household frame is $n_{SF} = C^*/C_{HH} = 150{,}289$, which is only slightly greater that the total sample for the **DF** design. The variance of the single-frame estimate will therefore be

$$V_{SF}(p_{HH}) = \frac{P(1-P)}{n_{SF}} = \frac{0.001(1-0.001)}{150{,}289} = 6.647\times 10^{-9}.$$

Cost-Equivalent Design Comparison:

Comparing the variances for RSA estimates from the **DF** and **SF** designs with $C^* = \$26M$, we have

$$R_V = \frac{V_{DF}^{(C-E)}(p_W)}{V_{SF}(p_{HH})} = 0.549.$$

implying that the variance for the **DF** design is about 45% lower than the cost-equivalent variance for the **SF** design.

EXAMPLE 2: [$\theta = C_A/C_{HH} = 10$]

Consider the same setting as above but where $\theta = C_A/C_{HH} = 10$; i.e., where the average cost for the police records stratum is *10 times greater* than for the household stratum (e.g., because it may be much more difficult to sample, recruit, and collect data from the sample obtained from police records). Here, the most cost-efficient allocation of the **DF** sample changes to $n_A^{(C-E)} = 420$ and $n_{HH}^{(C-E)} = 146{,}086$, and the variance ratio is $R_v = 0.556$, implying a 43% lower variance by the **DF** design.

1. An important factor in the much higher average unit cost for the police records stratum is the need to broaden the search for RSA cases beyond those persons reporting assaults/attempts to com-

mit rape (e.g., to also include aggravated assaults by a male on a female) so that, we note that the following changes in R_v when P_A is smaller:

P_A	R_v
0.60	0.709
0.50	0.768
0.40	0.825
0.30	0.879
0.20	0.930

These findings indicate that even at lower concentrations and substantially higher average unit costs for this administrative source, the dual-frame approach produces reasonable gains over a cost-equivalent single-frame approach.

2. I have produced a wider range of findings for all of the statistical and process indicators just computed to more broadly illustrate comparative results for the dual-frame approach versus a cost-equivalent single-frame approach when police records are the administrative frame source for the dual frame.

SOME FINAL THOUGHTS

Admittedly, the utility of the comparative findings in this document is somewhat limited by several simplifying assumptions I have made, particularly by (i) the use of a contrived two-stratum framework for the two overlapping frames of the dual-frame by screening out target population members from one frame in sampling the other, and (ii) the assumption of SRSWR sampling instead of further stratified multistage cluster sampling in each stratum,[4] and (iii) considering only effects on sampling error instead of also including effects arising from other nonsampling sources errors such as nonresponse and measurement. Nonetheless, I believe that these preliminary findings strongly suggest that it would be worthwhile for BJS to more closely investigate the feasibility of using a dual-frame approach for estimating rates of RSA, particularly if these estimates are obtained from an independent RSA victimization survey as recommended by the panel. Finally, the panel's suggestions accompanying a further investigation of the dual-frame might be to incorporate more realistic elements overlooked by my simplifying assumptions above.

[4]Kalsbeek, Spencer, and House (2013) provide more information on the potential efficiency reductions expected from relaxing this assumption.

Appendix F

Biographical Sketches of Panel Members and Staff

Candace Kruttschnitt *(Cochair)* is a professor in the Department of Sociology at the University of Toronto. Previously, she was chair of the department of sociology at the University of Minnesota. She has worked and published extensively on the subjects of female offending and victimization. Her current research focuses on the effects of confinement on offenders in different political and cultural contexts (the United States, Britain, and the Netherlands). She was appointed a visiting professor by the Royal Netherlands Academy of Arts and Science in 2010, and was awarded the Distinguished Scholar Award in 2011 by the American Society of Criminology's Division on Women and Crime. She is the elected president (for 2015) of the American Society of Criminology. She has a B.A. in criminology from the University of California, Berkeley, and an M.A., a M.Phil., and a Ph.D. in sociology from Yale University.

William D. Kalsbeek *(Cochair)* is a professor in the Department of Biostatistics and past director of the Survey Research Unit at the University of North Carolina at Chapel Hill. His research interests include the practice, teaching, and study of health survey research methods, with particular emphasis on the joint roles of costs and statistical efficacy in survey design, sampling of elusive populations, and survey nonresponse. He has served as sampling statistician or adviser on many of the major national surveys in the United States and has designed samples for several international health surveys, including ones in Egypt, Honduras, Indonesia, Jordan, Russia, and Somalia. He is a fellow of the American Statistical Association. He has a

B.A. in mathematics from Northwestern College and a M.P.H. and a Ph.D. in biostatistics from the University of Michigan.

Paul P. Biemer is a distinguished fellow in statistics at RTI International and associate director for survey research and director of the certificate program in survey methodology at the Odum Institute for Research in Social Sciences at the University of North Carolina at Chapel Hill. Previously at RTI, he served as director of the survey methods program and director of the Center for Survey Methods and Research. His work in survey methodology and statistics includes developing methodologies for using computer audio-recorded interviewing, using latent class analysis as a survey error evaluation tool, and applying continuous quality improvement to the coding of industry and occupation question responses. He has a B.S. in mathematics, an M.S. in statistics, and a Ph.D. in statistics from Texas A&M University.

John Boyle is senior vice president and survey research line of business lead for ICF International. Previously, he was executive vice president of Abt SRBI, a senior partner of SRBI, and senior vice president of Louis Harris and Associates. His study areas include epidemiology, health care utilization and outcomes, violence and post-traumatic stress disorder, service quality assessment, program evaluation, and policy analysis. He has worked extensively in the design, execution, and analysis of surveys related to sexual assault, victimization and abuse, including both military and national civilian surveys. He directed the National Violence Against Women Survey and the National College Women Sexual Victimization Survey, among others. He has a Ph.D. from Columbia University.

Bonnie S. Fisher is a professor in the school of criminal justice at the University of Cincinnati. Previously, she was on the faculty of the Department of City and Regional Planning at Ohio State University, a visiting scholar in the Division of Prevention and Community Research at Yale University School of Medicine, and a visiting professor in the Department of Sociology at John Jay College of Criminal Justice. She was the principal investigator for major research projects on the victimization of college students, the sexual victimization of college women, violence against college women, and campus-level responses to a report of sexual assault. Her research interests include sexual violence against women, repeat victimization, fear of crime, the measurement of victimization, injury detection of rape victims, and the court's use of digital images in the prosecution of rape cases. She has a Ph.D. in political science from Northwestern University.

Karen Heimer is professor of sociology and women's studies at the University of Iowa. She is also affiliated with the Public Policy Center at the

University of Iowa, where she served as Director of Social Science Research in the past. Her recent work includes studies of violence against women and minorities over time, women's offending and victimization across cities and metropolitan statistical areas over time, race and gender differences in imprisonment over time, and juvenile delinquency. Heimer is Vice President of the American Society of Criminology in 2014 and has served as an executive council member of the American Society of Criminology in the past. She also served a term as Chair of the Crime, Law and Deviance Section of the American Sociological Association. She received her M.A. in psychology and Ph.D. in sociology from the University of Wisconsin–Madison.

Carol C. House (*Study Director*) is a senior program officer for the Committee on National Statistics. She previously served as study director for the project that resulted in *Measuring What We Spend: Toward a New Consumer Expenditure Survey*, and was a rapporteur for *Research Opportunities Concerning the Causes and Consequences of Child Food Insecurity and Hunger: A Workshop Summary*. Prior to her work the National Research Council, she held several positions at the National Agricultural Statistics Service of the U.S. Department of Agriculture, including deputy administrator for programs and products, associate administrator, director of research and development, and director of survey management. She also served as chair of the Agricultural Statistics Board. She has provided statistical consulting on sample surveys in China, Colombia, the Dominican Republic, and Poland. She is a fellow of the American Statistical Association and an elected member of the International Statistical Institute. Her graduate training was in mathematics at the University of Maryland.

Linda Ledray is director of the SANE-SART Resource Service and founder of the Minneapolis-based Sexual Assault Resource Service (SARS), which she directed for more than 30 years. SARS provides forensic-medical services to sexual assault survivors at seven hospital sites, and was one of the first Sexual Assault Nurse Examiner (SANE) programs in the country. She has helped develop and implement SANE programs across the United States and in numerous other countries. She was a founding member of the certification board of the International Association of Forensic Nurses, and serves on the editorial board of its journal. Her articles have appeared in both scholarly and popular journals, and she has appeared on CNN, Arthur Frommer's Almanac, and CBS This Morning. She has a B.S. in nursing and an M.A. in community and mental health nursing, both from the University of Washington, and an M.A. in psychology and a Ph.D. in clinical psychology and personality research from the University of Minnesota.

Colin Loftin is a professor in the School of Criminal Justice at the University and codirector of the Violence Research Group, a research collaboration with colleagues at the University at Albany and the University of Maryland. The Violence Research Group conducts research on the causes and consequences of interpersonal violence In four broad areas: understanding violence as a social process extending beyond individual action; improving the quality of data on the incidence and nature of crime; the design and evaluation of violence prevention policies; and the investigation of population risk factors for violence. He has a B.A. (with honors), an M.A., and a Ph.D. from the University of North Carolina at Chapel Hill.

Ruth D. Peterson is emerita professor of sociology, former distinguished professor of social and behavioral science, and former director of the Criminal Justice Research Center, and she coordinates the Racial Democracy, Crime and Justice Network, all at Ohio State University. Her research focuses on community conditions and crime, racial and ethnic inequality in patterns of crime, and the consequences of criminal justice policies for racially and ethnically distinct communities. Her current work focuses on attempts to explain how and why patterns of neighborhood crime vary across communities of different colors and explores how residential and nonresidential neighborhoods that individuals traverse during the course of their daily activities differentially influence their participation in crime, drug use, and other problem behaviors. She has a Ph.D. in sociology from the University of Wisconsin-Madison.

Nora Cate Schaeffer is Sewell Bascom professor of sociology at the University of Wisconsin– Madison, where she also serves as faculty director of the university's Survey Center, teaches courses in survey research methods, and conducts research on questionnaire design and interaction during survey interviews. She currently serves as member of the Advisory Board of the *Public Opinion Quarterly* of the American Association for Public Opinion Research and of the Board of Overseers of the General Social Survey. She has also served on advisory and technical committees for the Survey of Income and Program Participation, the National Longitudinal Survey of Youth, the National Science Foundation, and the governing council of the American Association for Public Opinion Research. She is a fellow of the American Statistical Association. She has a Ph.D. from the University of Chicago.

Tom W. Smith is a survey methodologist and director of the General Social Survey at NORC at the University of Chicago. He is frequently consulted and has spoken publicly on a wide range of topics, including American sexual behavior, intergroup relations, confidence in institutions, happiness, religion, guns, and voter behavior He has been awarded the Worcester Prize of the *International Journal of Public Opinion Research*; the Innovators Award and the Award for Exceptionally Distinguished Achievement of the American Association for Public Opinion Research ; and the award for distinguished contributions to sociology of the Eastern Sociological Society. He has a B.A. in history and political science and an M.A. in history from Pennsylvania State University and a Ph.D. in American history from the University of Chicago.

Bruce D. Spencer is a professor and former department chair of statistics and a faculty fellow of the Institute for Policy Research at Northwestern University. His interests span the disciplines of statistics and public policy with a special focus on the design and evaluation of large-scale statistical data programs. He is currently studying the costs and benefits of alternative designs for the 2020 census, how to estimate the accuracy of verdicts in criminal cases when the truth is unknown, and how to draw inferences about networks from samples. Previously, he directed the Methodology Research Center of NORC at the University of Chicago. He received the Palmer O. Johnson Memorial Award from the American Educational Research Association and is an elected fellow of the American Statistical Association. He has a Ph.D. from Yale University.

COMMITTEE ON NATIONAL STATISTICS

The Committee on National Statistics was established in 1972 at the National Academies to improve the statistical methods and information on which public policy decisions are based. The committee carries out studies, workshops, and other activities to foster better measures and fuller understanding of the economy, the environment, public health, crime, education, immigration, poverty, welfare, and other public policy issues, it also evaluates ongoing statistical programs and tracks the statistical policy and coordinating activities of the federal government, serving a unique role at the intersection of statistics and public policy, The committee's work is supported by a consortium of federal agencies through a National Science Foundation grant.